SpringerBriefs in Eth

More information about this series at http://www.springer.com/series/10184

Barbro Fröding · Walter Osika

Neuroenhancement: How Mental Training and Meditation Can Promote Epistemic Virtue

Barbro Fröding
Department of Philosophy and History
Royal Institute of Technology (KTH)
Stockholm
Sweden

Walter Osika
Department of Clinical Neuroscience, Center for Psychiatry Research
Karolinska Institutet
Stockholm
Sweden

ISSN 2211-8101 ISSN 2211-811X (electronic)
SpringerBriefs in Ethics
ISBN 978-3-319-23516-5 ISBN 978-3-319-23517-2 (eBook)
DOI 10.1007/978-3-319-23517-2

Library of Congress Control Number: 2015948153

Springer Cham Heidelberg New York Dordrecht London

Springer International Publishing AG Switzerland is part of Springer Science+Business Media (www.springer.com)

For Cecilia and Simon

Acknowledgments

I am greatly indebted to a number of people for helping me to develop and refine the ideas presented in this book. Without their insights, suggestions, comments, and constant encouragement, I would not have been able to write this. A very special thanks goes to Professor Martin Peterson, Professor Sven Ove Hansson, Professor Wolf Singer, Professor Roger Crisp, and Professor Julian Savulescu, as well as to the colleagues and friends at the Oxford Uehiro Centre for Practical Ethics. All errors that remain are of course my own.

Barbro Fröding

There are many people who have inspired me to dare to co-write this book. First, I want to thank my co-author Barbro Fröding, who invited me to be her co-traveler on this expedition. A very special thanks goes to my long-time supervisor Prof. Arthur Zajonc (Amherst College, Mind and Life) Marie Ryd Ph.D., and Cecilia Stenfors Ph.D. for inspiring intellectual conversations, as well as to the colleagues and friends at Center for Social Sustainability, Department for Clinical Neuroscience, Karolinska Institutet and the Stress Clinic, Stockholm, Sweden. All errors that remain are of course my own.

Walter Osika

Contents

Chapter 1
Introduction

Where is the Life we have lost in living?
Where is the wisdom we have lost in knowledge?
Where is the knowledge we have lost in information?
From 'The Rock' by T.S Eliot, 1934

Abstract This chapter provides a background and a schematic overview of the book. In this volume we argue that meditation enables us to influence some aspects of our biological make-up and could, for example, boost our cognitive flexibility as well as our ability (and propensity) to act compassionately. Then we proceed to seek to connect a number of such changes to an improved capacity for instilling and maintaining a range of character traits (primarily epistemic virtues) as identified by Aristotle and some modern virtue epistemologists. Cultivating the virtues is of course beneficial for the individual but it seems likely that it also has a positive effect on the surrounding society and their fellow citizens.

Keywords Virtue · Decision-making · Information · Meditation · Mental training · Cognitive flexibility · Compassion · Behaviour

1.1 Introduction

Recent findings in the natural sciences have confirmed that it is possible for humans to achieve measurable structural and functional changes in the brain through various life-style practices. Examples range from aerobic exercise, juggling, music training, learning a foreign language to the training course to become a licensed taxi driver in London.[1] This means that the adult brain is more plastic than previously thought and, further, it seems that these changes are lasting and generalizable.

[1]Draganski, B., Gaser, C., Busch, V., Schuierer, G., Bogdahn, U., & May, A. (2004). Neuroplasticity: changes in grey matter induced by training. *Nature*, *427*(6972), 311–312. Pinho, A. L., de Manzano, Ö., Fransson, P., Eriksson, H., & Ullén, F. (2014). Connecting to Create: Expertise in Musical Improvisation Is Associated with Increased Functional Connectivity between Premotor and Prefrontal Areas. *The Journal of Neuroscience*, *34*(18), 6156–6163. Mårtensson, J., Eriksson, J., Bodammer, N. C., Lindgren, M., Johansson, M., Nyberg, L., & Lövdén, M. (2012). Growth of language-related brain areas after foreign language learning. *Neuroimage*, *63*(1), 240–244. Woollett, K., & Maguire, E. A. (2011). Acquiring "The Knowledge" of London's layout drives structural brain changes. *Current biology*, *21*(24), 2109–2114.

B. Fröding and W. Osika, *Neuroenhancement: How Mental Training and Meditation Can Promote Epistemic Virtue*, SpringerBriefs in Ethics,
DOI 10.1007/978-3-319-23517-2_1

Hence, we are able to support the development of neurophysiological connections and functions—and consequently new behavior—through the activities we chose to engage in and the habits we cultivate. In this book we discuss how such changes can be achieved through regular mental training in the form of meditation. In particular, we look at the beneficial effects of Attention Training Meditation, Open Monitoring (as in e.g. Mindfulness training) and Compassion Training.

We will argue that meditation enables us to influence some aspects of our biological make-up and could, for example, boost our cognitive flexibility as well as our ability (and propensity) to act compassionately. Then we proceed to seek to connect a number of such changes to an improved capacity for instilling and maintaining a range of character traits (primarily epistemic virtues) as identified by Aristotle and some modern virtue epistemologists. The reason for why this would be good and desirable is that agents who are successful in instilling such and other virtues plausibly can be taken to lead happier lives (from an all things considered perspective) than those who fail to do so. Cultivating the virtues is of course beneficial for the individual but it seems likely that it also has a positive effect on the surrounding society and fellow citizens.[2]

The research we have used here is part of the increasing number of well-designed studies where the integration of *first person* accounts of meditation practice experiences (diaries, self-assessment scales) with second person (e.g. interviews) and third person perspective (e.g. neuroimaging) has been the objective.[3] Notably, the argument here is not that meditation techniques somehow automatically allow us to access and/or tap into capacities or knowledge we already have.

We are well aware that a detailed understanding of exactly how behavior changes due to different practices is still far away. What we have aimed for on this volume is to provide a conceptual take on mental training and meditation from a neuroscience perspective. In this we have made some assumptions including (i) that it is possible to impact behavioral outcomes with mental training and (ii) that this could be of importance if one wants to lead a virtuous life.

The account of virtue ethics that is to follow is very brief and focuses primarily on the epistemic virtues. None the less, it will hopefully provide some theoretical background for the more general discussion in this book. To be clear, this is not intended as an authoritative, or exegetic, reading of Aristotle. Rather, our aspiration is to suggest that many of the ideas in the Nicomachean Ethics (Aristotle's central work on ethics) make for a highly useful approach to modern moral problems.

---

[2]Note that while this position evidently is based on virtue ethics and Aristotle's account this underlying theory will not be defended in this article for space reasons.

[3]Varela F, Thompson E, Rosch E. (1991) *The Embodied Mind: Cognitive Science and Human Experience* MIT Press. Gallagher S, Zahavi D. (2012) *The Phenomenological Mind.* Routledge, Oxon.

1.2 A New Situation

As a result of rapidly advancing technological developments society is becoming increasingly complex and cognitively demanding. Consider, for example, the torrent of information and disinformation that many of us face on a daily basis. To handle this information in a manner that is conducive to responsible decisions is proving a great challenge for many people. One explanation is found in the way the human brain functions. Humans, in general, are for biological reasons not very good at responding to information. This inability extends both 'the gathering of information phase' and to 'the drawing conclusions and acting on them phase'.

The combination of the increasing flow of information in society, on the one hand, and our relatively meagre cognitive capacities and emotional repertoire, on the other, effectively stops us from living as good and happy life as we could. Our collective failure to process the information and then make responsible decisions based on our conclusions[4] has disastrous consequences. We harm ourselves, each other and the planet.

This new situation places novel, or at last increased, cognitive as well as emotional demands on people. The point of departure for this work is the thought that to meet such demands many of us would benefit from improved cognitive capacities, better emotional regulation and more well-grounded moral choices. Contrary to popular belief most humans are severely challenged in the area of rational decision-making, something which frequently has negative consequences. Numerous scientific studies in the fields of, for example, neurology and neuropsychology have shown that information overload and stress negatively influence cognitive capacities such as memory (both process and long-term), risk assessment and epistemic deference.[5]

Another effect of stress is that we become more rigid and stuck in our old ways even when we ought to change. It appears that stress plays havoc with our capacity for emotional regulation and makes us fearful of things simply in virtue of them being new (as opposed to actually dangerous or less good than current practices).[6] One of the causes of maladaptive coping with stress, which has been highlighted lately, is "experiential avoidance".[7] This can be described as a kind of phobia for

---

[4]Iyengar, Sheena. *The art of choosing*. Twelve, 2010.

[5]Qin S, Hermans EJ, van Marle HJ, Luo J, Fernandez G (2009) Acute psychological stress reduces working memory-related activity in the dorsolateral prefrontal cortex. *Biological Psychiatry* 66:25–32. Arnsten, A. F. (2009). Stress signalling pathways that impair prefrontal cortex structure and function. *Nat Rev Neurosci, 10*(6), 410–422.

[6]Golkar, A., Johansson, E., Kasahara, M., Osika, W., Perski, A., & Savic, I. (2014). The influence of work-related chronic stress on the regulation of emotion and on functional connectivity in the brain. *PloS one*, *9*(9), e104550. For a discussion on 'status quo bias' and its consequences see Chap. 3.

[7]Hayes, S. C., Wilson, K. G., Gifford, E. V., Follette, V. M., & Strosahl, K. (1996). Experiential avoidance and behavioral disorders: A functional dimensional approach to diagnosis and treatment. *Journal of consulting and clinical psychology*, *64*(6), 1152.

internal experiences such as thoughts, feelings, memories and physical sensations. Psychological research shows that it is not negative thoughts, emotions, and sensations that are problematic *per se*, it is rather how one responds to them that can cause the difficulties.[8] If one is unable to harbor, or at least reluctant to experience, uncomfortable thoughts and feelings, that might increase the risk of suboptimal stress management which, in turn, is thought to be linked to other forms of problems. Everyday examples of such problems include: the putting off important tasks because of the discomfort they evoke, or spending so much time attempting to avoid discomfort, that you have little time for anyone or anything else in your life. A significant impact of experiential avoidance is its potential to disrupt and interfere with important, valued aspects[9] of an individual's life. From a virtue ethics perspective experiential avoidance can be seen as particularly problematic when it occurs at the expense of a person's deeply held values. It could be taken as evidence that the person has not instilled the virtues and therefore their desire (as opposed to their reason) is controlling the decision-making and they attach pleasure to the wrong things and actions. Stress can also in itself increase the tendency to experiential avoidance, which could create a vicious circle of avoidance, increased stress and so forth. (We return to some practical examples below in Sect. 1.4 and in Chap. 3).

Moral values are of importance in decision-making—we seek and categorize information depending on our worldview and our values. An increased awareness of our own pre-understanding (including bias) might have a positive effect on our ability to handle complex and sometime contradictory information.[10] For example, it could enhance our ability to integrate fragmented, confusing facts in our decision-making process, and thereby make us more prone to act in a more well-informed, "all things considered" way.

Broadly speaking 'good decision-making' could, at least in the context of this book, be understood as 'responsible, reflected and rational decision-making', which then would imply that we need to foster a more pro-social behavior and improve our general life-skills. Some concrete examples of pro-social behavior could include a raised sense of responsibility for matters that fall in the collective domain,[11]

---

[8]Wegner, Daniel M.; Schneider, David J.; Carter, Samuel R.; White, Teri L. (1987). "Paradoxical effects of thought suppression". *Journal of Personality and Social Psychology* 53(1): 5–13.

[9]Valued direction is a concept that is used in counselling and psychological treatment, in e.g. acceptance and commitment therapy (ACT). The client/patient is encouraged to make ratings of what is important to him/her and, then, rate how satisfied one is with the current state of being in these areas, and then formulate ones intentions, ones valued direction in life. Eifert, G. H., & Forsyth, J. P. (2008). *The mindfulness and acceptance workbook for anxiety: A guide to breaking free from anxiety, phobias, and worry using acceptance and commitment therapy*. New Harbinger Publications.

[10]We define a bias here as a form of heuristic or shortcut that the human brain is prone to when engaging in e.g. decision-making, general assessment of events, ranking how important events/facts are and what to pay attention to in a situation.

[11]Avoiding the tragedy of the commons scenarios as described by e.g. Hardin, G. (1968). The tragedy of the commons. *Science*, *162*(3859), 1243–1248.

a willingness to adopt a sustainable life-style, and other 'living together' aspects as broadly conceived of.

Arguably, the fact that most of us are left wanting when it comes to good decision-making has a negative impact on our overall quality of life. It can hardly be denied that human wellbeing and quality of life would improve radically were people in general to become better at making decisions—to behave (more) responsibly and consider the interests of others as well as the long-term consequences of their actions.

Indeed, our inability to achieve lasting collaborations, assess risks and comprehend the long-term consequences of our actions has highly unpalatable results. For some concrete examples, consider the lack of constructive co-operation on how to address global warming, how to manage the political and social fallout of the financial crisis and how to avoid, or at least alleviate, armed conflicts. It would appear that we, both as a collective and as individuals, can ill afford poor decision-making and given the pressing nature and potentially catastrophic consequences of such failure it is of interest to further explore cognitive enhancement.

Unfortunately, high cognitive flexibility does not automatically entail moral behavior and good decision-making (in the impartial, responsible, pro-social way we have in mind). Consider for example the darkly creative individual who uses their capacity to manipulate others or an individual who turns cold and calculating as a result of an improved capacity to detach and be "objective". We can also imagine that being highly creative could correlate with a tendency to being risk prone and not pay proper attention to prudence and moderation.[12] To manage such bleak scenarios we need a moral framework to harness and anchor these capacities, and as we will show in this book virtue ethics is especially well suited to help us with this.[13]

All that said, it could of course be the case that an individual who has managed to heighten her cognitive flexibility through a method involving, for example, Compassion Training might also have developed a deeper understanding of morality and the reasons for which one would choose to live in line with the moral virtues. It would, however, be imprudent to assume that such positive side-effects would occur automatically or in all individuals.

---

[12]Kyaga, S., Landén, M., Boman, M., Hultman, C. M., Långström, N., & Lichtenstein, P. (2013). Mental illness, suicide and creativity: 40-year prospective total population study. *Journal of psychiatric research*, *47*(1), 83–90.

[13]The focus of this book is cognitive enhancement (as opposed to physical and moral enhancements) and how that could be achieved through committing to meditation and virtue ethics. Note however, that some philosophers have argued that in order to handle the potential dangers of cognitive enhancement humans need moral enhancement and, further, that this would be best brought about not through virtue ethics but rather by pharmaceutical drugs and or hormones. For an interesting account see Douglas, T. (2008). Moral enhancement. *Journal of applied philosophy*, *25*(3), 228–245. Persson, I., & Savulescu, J. (2008). The perils of cognitive enhancement and the urgent imperative to enhance the moral character of humanity. *Journal of Applied Philosophy*, *25* (3), 162–177. See also Chap. 6 for a discussion.

1.3 From Bad to Worse

Every year, in connection with the Davos meetings, The World Economic Forum issues a Global Risks Report listing the biggest threats to our future prosperity and security. Major risks identified in recent reports include; income disparity, extreme weather events, fiscal crisis and technology risks like cyberattacks and critical information infrastructure breakdown.[14] These are scenarios considered capable of causing systemic shock on a global scale. In addition, the reports bring up the threat of the 'x-factor', i.e. events that cannot be foreseen and the consequences of which are unknown.

Some key risks that have been identified with regards to human health are a sense of hubris and an inflated belief in the progress of medical technology and pharmaceutical drugs. It appears that many of us subscribe to an overly optimistic view of both the speed of the development, and the efficiency, of the novel treatments. For a concrete example consider the current use of antibiotics which is resulting in bacteria mutation and increased resistance to available treatments as well as unforeseen side effects in other species and entire eco-systems.[15] It would seem that our belief in the ever advancing technology and the discoveries in the natural sciences can occasionally lead us to believe that we are in a better position than we really are with regards to e.g. pandemics and chronic illnesses. The Global Risk Report paints a bleak picture: while we might be able to diminish some of the negative consequences it appears unlikely that we can avoid all of them.[16]

1.4 How to Handle Matters Better: Skills Required to Manage Risks

On a slightly more positive note, the Forum also releases a Special Report. This report looks at how countries could become *better prepared* to face risks of a global nature, i.e. risks that seem to be beyond their control. The authors observe: "One possible approach rests with "systems thinking" and applying the concept of resilience to countries". (The World Economic Forum's *Global Risks 2013* Report). To manage and negotiate this type of highly challenging situations requires a wide

[14]http://www.weforum.org/reports/global-risks-2014-report.

[15]See e.g. Laxminarayan, R., Duse, A., Wattal, C., Zaidi, A. K., Wertheim, H. F., Sumpradit, N., … & Cars, O. (2013). Antibiotic resistance—the need for global solutions. *The Lancet infectious diseases*, *13*(12), 1057–1098. Deblonde, T., & Hartemann, P. (2013). Environmental impact of medical prescriptions: assessing the risks and hazards of persistence, bioaccumulation and toxicity of pharmaceuticals. *Public health*, *127*(4), 312–317.

[16]But, as shown here, they can be mitigated. See Chaps. 2 and 6 for a longer discussion on how a more accurate sense of self-assessment (i.e. that one that stays clear both of hubris as well as misplaced concerns and insecurity) can be cultivated through meditation techniques and compassion training.

range of cognitive capacities e.g. cognitive flexibility, creativity, intellectual courage, honesty, wisdom, impartiality, commitment to fairness and a sense of reciprocity.[17] Plausibly, we could improve on the current state of affairs and limit the impact both on people's quality of life and on the planet but it requires collaboration, commitment, research and the allocation of financial resources. It would appear that we have ample reason to seek to improve our decision-making skills.

1.5 Can We Acquire the Skills?

At the beginning of this chapter we suggested that humans as a collective need to start making better decisions. We sketched a brief account of what we believe to be a rather pressing situation; a combination of a changing society and research confirming that our cognitive capacities are more limited, and our bias more substantive, than what was previously known. Taken together, all these factors conspire to undermine the quality of our decision-making which, in turn, negatively impacts our quality of life and might even pose a threat to our future. But needs aside—how capable are we?[18]

1.5.1 Half-Willing but Unstable?

On a general level most people can choose to behave in ways (i.e. make better decisions as broadly conceived of) that make their lives go slightly better than what otherwise might be the case. The traditional recipe for promoting the desired behaviour, especially in the moral domain, relies heavily on the application of unfaltering will-power, commitment and self-discipline. In other words, the assumption is that refraining from doing X and Y will not necessarily get easier with time, at least not in the sense that one becomes less sensitive to the pull of temptation. It is of course possible that one gets better at controlling or ignoring said temptation but plausibly such control would be related to fear of negative consequences such as punishment or social stigma.[19]

---

[17]Missimer, M. (2013). *The social dimension of strategic sustainable development*. Licentiate Dissertation, Blekinge Institute of Technology. Ostrom, E. (1990). *Governing the commons: The evolution of institutions for collective action*. Cambridge University press.

[18]Free will and moral responsibility falls outside the scope of this volume but for a very interesting account of an agency cultivation model (of how holding someone morally accountable for their actions and the effects of practices matters when it comes to cultivating moral agency) see Vargas, M. (2013). *Building better beings: A theory of moral responsibility*. Oxford University Press.

[19]These are the type of people that Aristotle referred to as continent. They know what they should do and most of the time they are able to act accordingly but since the virtues are not properly instilled they are not reliable decision-makers. See Chap. 5.

Looking around the world today one could indeed be forgiven for entertaining, or at least not ruling out, the possibility that human nature is flawed in a way that makes us selfish opportunist and by and large unable to instil stable traits which can guide our actions (hence the rather elaborate penal systems). Whether or not humans can have stable character traits or if our actions tend to depend (more) on the circumstances or the situation has long been subject to a heated debate within the philosophical community.[20] More recently, however, research in the fields of neuropsychology, moral psychology and even behavioural economics have added fuel to the fire.

1.5.2 Systematically Enslaved?

Daniel Kahneman and Adam Twersky have produced seminal work on human decision-making and the effects of systemic bias.[21] In a series of experiments they showed that the human decision-making process is flawed by a proneness to a number of systemic bias which have a rather negative impact on the quality of the outcome. Examples of such effects include, loss aversion, sunk cost, lacking capacity for risk assessment, over optimism, framing effects and substitution (i.e. replacing the tricky problem at hand with a simple one pretending that they are analogous or at least similar enough). Before we turn to look at some of Kahneman's and Twersky's key insights and discuss how they are relevant for the ideas advocated here, a few brief comments on the functioning of the human brain might be of interest.

The human brain has a strong dislike for cognitive dissonance. In fact, this is such a stressful state that we are prepared (consciously or sub-consciously) to change our narrative and general assessment of situations in order to harmonize

[20]For both sides of the discussion see e.g. Harman. (1999). Moral philosophy meets social psychology: virtue ethics and the fundamental attribution error. *Proceedings of the Aristotelian Society* 99: 315–331. Darley, J. M., & Batson, C. D. (1973). "From Jerusalem to Jericho": A study of situational and dispositional variables in helping behavior. *Journal of Personality and Social Psychology*, *27*(1), 100. Doris, J. M. (2002). *Lack of character: Personality and moral behavior*. Cambridge University Press. Haidt, J., Seder, J. P., & Kesebir, S. (2008). Hive psychology, happiness, and public policy. *The Journal of Legal Studies*, *37*(S2), S133–S156. Hursthouse. 1991. Virtue theory and abortion. *Philosophy and Public Affairs* 20(3): 223–246. Nussbaum, M. (1986). *The fragility of goodness: luck and ethics in Greek tragedy and philosophy*. Cambridge University Press. Milgram, S. (1963). Behavioral study of obedience. *The Journal of Abnormal and Social Psychology*, *67*(4), 371. Hobbes, T. (1969). *Leviathan, 1651*. Scholar Press.

[21]For the purpose of this discussion, we have chosen to use the umbrella term "systemic bias" describing the inherent tendency of a (mental) process to primed by biological and social/environmental factors, which subsequently influences behavior and decision making. Unfortunately the agent tends to be unaware of the nature and magnitude of such bias. See e.g. Tversky, A., & Kahneman, D. (1973). Availability: A heuristic for judging frequency and probability. *Cognitive psychology*, *5*(2), 207–232. Kahneman, D., & Tversky, A. (1996). On the reality of cognitive illusions.

various data input. This can involve a radical revision of views and beliefs that most of us would like to think we are deeply attached to e.g. political values and voting patterns.[22] In addition to this, there is also an inherent tendency in the brain to identify meaningful patterns in random data (*apophenia*) and to perceive vague and random images and sound as significant (*pareidolia*), not least human faces.[23]

The human brain is restricted by ancient behavioural patterns which are mirrored neuro-physiologically in more or less evolutionary segregated brain networks. One consequence of this is that humans have two main ways to reach a solution when we are faced with a situation, or a problem. In the literature the two ways are often referred to as System 1 and System 2. Typically, System 1 thinking is fast, lacking in reflection, instinctive, gut-driven, automatic, emotional, stereotyping and sub-conscious while System 2 thinking tends to be slow, effortful, infrequent, logical, calculating, conscious etc. It would however, be unfair to blame all the bias on the much vilified System 1—notably *both* systems generate systemic bias but often of different kinds. Consider, for example, how System 2 thinking can derail into agonized, unproductive rumination and action paralysis.

A common scenario is that System 1 produces a conclusion and then System 2 takes over and come up with reasons that justify the conclusions issued by System 1. But the problem is that many of us think that it works the other way around. Perhaps understandably most of us like to believe that we ponder and mull and reason very rationally (i.e. use System 2) and then reach a well-founded and well-considered conclusion. This would fit better with our self-image and the overall narrative we have constructed about ourselves and the world. Furthermore, many of us seem to believe that in situations where we instantaneously feel we know what is right it follows that we *also* have good reasons to believe the correctness of our conclusion. Indeed, the fact that we 'knew immediately' is taken as evidence of the correctness of the conclusion. While it is of course possible that we are right (by sheer coincidence and luck) this type of approach appears to be an unreliable method since it involves a minimum of reasoning and deliberation.

In a series of experiment Rand et al. showed that subjects who reach their decisions more quickly are more cooperative in economic games (and we can suppose that System 1 is more in charge here).[24] When the subjects were stressed to decide quickly, their contributions increased, but when they were instructed to reflect and were forced to decide slowly, their contributions decreased. If they were

---

[22]For an interesting study involving people's political values on the combination of in-group preferences with the above-average effect in American voters see Eriksson, K., & Funcke, A. (2012). American political ingroup bias and the above-average effect. *Available at SSRN 2168264*.

[23]Hadjikhani, N., Kveraga, K., Naik, P., & Ahlfors, S. P. (2009). Early (N170) activation of face-specific cortex by face-like objects. *Neuroreport*, *20*(4), 403. Voss, J. L., Federmeier, K. D., & Paller, K. A. (2011). The potato chip really does look like Elvis! Neural hallmarks of conceptual processing associated with finding novel shapes subjectively meaningful. *Cerebral Cortex*, bhr315.

[24]Rand, D. G., Greene, J. D., & Nowak, M. A. (2012). Spontaneous giving and calculated greed. *Nature*, *489*(7416), 427–430.

primed to trust their *intuitions* it increased contributions compared with priming that promoted greater reflection. The authors proposed that cooperation is intuitive, and one interpretation would be that System 1 in this case enhanced what might be, broadly, considered as virtuous behaviour. This finding has however been questioned and, looking at the existing literature, the 'intuitive-cooperation' effect appears to be rather variable.[25]

From a neuroscience perspective, it has recently been shown that "lower", subcortical structures (i.e. the amygdalae) also are very important for emotional decision-making (which could be said to represent System 1). For example, the amygdalae react immediately when we are presented to unjust scenarios and this response has been interpreted as a *moral reaction*.[26] Interestingly, the same study showed that treatment with a common tranquilizer (benzodiazepine) decreased the rejection rate concomitantly with a diminished amygdala response to unfair proposals, and this in spite of an unchanged feeling of unfairness. So the result is that the person still notices that the offer is unfair but as the reactions are attenuated s/he accepts to be conned. One could speculate that treating whole populations with tranquilizers could decrease their drive to protest against unfair societal circumstances.[27]

When people state their willingness to pay for something, the amount usually differs from the behavior in a real purchase situation. The discrepancy between a hypothetical answer and the real act is called *hypothetical bias*. Gospic et al.[28] investigated neural processes of hypothetical bias regarding monetary donations to public goods using fMRI. Real decisions actually activated amygdala more than hypothetical decisions. This was observed for both accepted and rejected proposals. The more the subjects accepted real donation proposals, the greater was the activity in rostral anterior cingulate cortex—a region known to control amygdala but also neural processing of the cost-benefit difference. The presentation of a charitable donation goal evoked an insula activity that predicted the later decision to donate. In conclusion, the neural mechanisms underlying real donation behavior are compatible with theories on hypothetical bias. The findings imply that the emotional

---

[25]Tinghög, G., Andersson, D., Bonn, C., Böttiger, H., Josephson, C., Lundgren, G., … & Johannesson, M. (2013). Intuition and cooperation reconsidered. *Nature*, *498*(7452), E1–E2. Verkoeijen, P. P., & Bouwmeester, S. (2014). Does intuition cause cooperation? *PloS one*, *9*(5), e96654.

[26]Here the "Ultimate Game" was used, an experimental game for studying the punishment of norm-violating behaviour. Gospic K, Mohlin E, Fransson P, Petrovic P, Johannesson M, et al. (2011) Limbic Justice—Amygdala Involvement in Immediate Rejection in the Ultimatum Game. *PLoS Biol* 9(5): e1001054.

[27]Vernon, M. (2011). Buddhism is the new opium of the people. *Te Guardian*, www.guardian.co.uk, *22*. Britton, W. B., Lindahl, J. R., Cahn, B. R., Davis, J. H., & Goldman, R. E. (2014). Awakening is not a metaphor: the effects of Buddhist meditation practices on basic wakefulness. *Annals of the New York Academy of Sciences*, *1307*(1), 64–81.

[28]Gospic, K., Sundberg, M., Maeder, J., Fransson, P., Petrovic, P., Isacsson, G., … & Ingvar, M. (2014). Altruism costs—the cheap signal from amygdala. *Social cognitive and affective neuroscience*, *9*(9), 1325–1332.

system has an important role in real decision-making as it signals what kind of immediate cost and reward an outcome is associated with.

Another interesting recent finding is that alcohol dependent individuals are more likely to endorse (what can be described as classical) utilitarian choices in personal moral dilemmas and, compared with controls and rate these choices as less difficult to make. Hierarchical regression models showed that poorer decoding of fear and disgust significantly predicted utilitarian biases in personal moral dilemmas, over and above alcohol consumption.[29] It is also increasingly evident that there are different triggers of moral disgust, with a differentiated flexibility and proneness to external justifications.[30]

In summary then; when the human brain strives to reach coherence and thus lower the stress-level, motivated reasoning[31] in combination with a set of other bias (e.g. status quo bias and confirmation bias) comes into play. Unsurprisingly, this has a rather negative impact on our decision-making. Over the next chapters we will develop our ideas and try to sketch how some of these tendencies can be off-set, or at least that we can become more aware of them, and act accordingly. We argue that (i) improved emotional regulation (including less experiential avoidance) and meta-awareness have a positive impact on cognitive flexibility, (ii) improved cognitive flexibility and meta awareness is likely to be highly conducive to the development of a set of epistemic virtues, e.g. by increasing the ability and willingness to notice our own biases, which (iii) could compensate for, or possibly even cancel out, some of our cognitive shortcomings and (iv) improve our decision-making (both the quality and propensity to act in accordance).

1.5.3 How Systemic Bias Make Us Vulnerable

It appears that we are overly confident and prone to think that we can foresee all possible outcomes of our actions. How such mistaken beliefs can develop into hubris (an inflated sense of self-importance and distorted reality) is a popular theme in e.g. leadership education and literature.[32] Hubris can be seen as an offshoot of

[29]Carmona-Perera, M., Clark, L., Young, L., Pérez-García, M., & Verdejo-García, A. (2014). Impaired Decoding of Fear and Disgust Predicts Utilitarian Moral Judgment in Alcohol-Dependent Individuals. *Alcoholism: Clinical and Experimental Research*, *38*(1), 179–185.

[30]In a recent review Russell & Giner-Sorolla present empirical evidence that moral disgust, in the context of bodily violations, is a relatively primitively appraised moral emotion compared to others such as anger, and also that it is less flexible and less prone to external justifications. They underscore the need to distinguish between the different consequences of moral emotion. Russell, P. S., & Giner-Sorolla, R. (2013). Bodily moral disgust: What it is, how it is different from anger, and why it is an unreasoned emotion. *Psychological bulletin*, *139*(2), 328.

[31]By this we mean decision-making based on un-reflected gut reactions NOT people's strong moral intuitions which would be well considered, stable and able to withstand the test of time.

[32]Romanowska J. (2014) Improving Leadership Through the Power of Words and Music. Doctoral dissertation. Karolinska Institutet, Stockholm, Sweden.

unbridled narcissism, a key force behind the desire for leadership and power. According to de Vries,[33] many leaders have strong tendencies towards such self-centeredness which, in turn, can lower their ability to step into those proverbial shoes of the other.[34] Further, there is substantial evidence that self-overconfidence reduces attempts to act and is associated with poor performance.[35]

Another unfortunate consequence of the combination of over confidence and systemic bias is that it leads us to grossly underestimate the complexity of the world and makes us (even more) vulnerable to what is called Black Swan events or 'the unknown unknowns'. A Black Swan is an event we could *not even imagine* prior to its coming about and the term is often used to point to the supreme fragility of any system of thought. The events are not only unpredicted but they have massive (often negative and always life altering) consequences. For some concrete examples of past events consider the personal computer, the internet, 9/11 and the fall of the Soviet Union.[36]

But even leaving such extreme scenarios aside our systemic bias also tend to make us struggle to assess both the likelihood and what might be a reasonable precaution strategy to avoid the 'known unknowns'. These are events that we can imagine but for various reasons fail to foresee and plan for. Consider for example the New Orleans flooding disaster and the human suffering and financial costs that still affect the region. Arguably, many of the consequences could have been mitigated and avoided with better preparation and a more developed levee/surge protection system.[37]

The problem is of course that hind-sight is always 20/20 and (barring a quantum like leap in technology) it is unlikely that we will be able to completely eradicate the unknown unknowns by foresight and preparation in the near future. If nothing else this would also require a massive re-allocation of resources which might be

---

[33]De Vries, M. F. K. (1990). The organizational fool: Balancing a leader's hubris. *Human Relations*, 43, 751–770.

[34]For how stress influences ethical decisions and is likely to reduce people's pro-social behaviours, and motivation to take others' interest into account see e.g. Jex, S., G. A. Adams, D. G. Bachrach, & S. Sorenson. (2003). The Impact of Situational Constraints, Role Stressors, and Commitment on Employee Altruism. *Journal of Occupational Health Psychology 8*(3), 171–180.

[35]For how stress has a negative influence on people's moral actions as well as on capacity for self-regulation and self-control see Selart, M., & Johansen, S. T. (2011). Ethical decision making in organizations: The role of leadership stress. Journal of Business Ethics, 99, 129–143. doi: 10.1007/s10551-010-0649-0.

[36]For a popularized discussion, see Taleb, N. N. (2010). *The Black Swan: The Impact of the Highly Improbable Fragility*. Random House. Note that Taleb has developed this line of thought further, and proposes that some systems can benefit from mistakes, faults, attacks, or failures, if the system has antifragile properties. Nassim Nicholas Taleb (2012). *Antifragile: Things That Gain from Disorder*. Random House. p. 430.

[37]Van Heerden, I. L. (2007). The failure of the New Orleans levee system following hurricane Katrina and the pathway forward. *Public Administration Review*, *67*(s1), 24–35. Pardue, J. H., Moe, W. M., McInnis, D., Thibodeaux, L. J., Valsaraj, K. T., Maciasz, E., … & Yuan, Q. Z. (2005). Chemical and microbiological parameters in New Orleans floodwater following Hurricane Katrina. *Environmental science & technology*, *39*(22), 8591–8599.

better spent on smaller but more likely events.[38] But while we may have to live with the Black Swans (and consequently strive to lead a life in an anti-fragile mode) we can get better at identifying and assessing risk in a more general sense. Research suggests that we can balance some of the negative effects of the system bias and consequently improve our overall situation and prospects, somewhat.

1.5.4 *Willing AND Able*

Fortunately all is not bad news on the science front. As previously mentioned there are other findings in neuroscience (e.g. in neurology and neuropsychology) which appear to offer us some consolation. In the last two decades numerous scientific studies have confirmed that we—by adopting a set of life-style habits—can achieve actual, physiological changes in the brain. This has been observed in a number of scientific studies as well as in concrete situations. A highly publicised example are the measurable changes in the brains of London taxi-drivers as they go through their training and similar results involving juggling and mental training aiming at developing specific abilities have been documented.

An increasing number of studies examining how regular meditation and other forms of mindfulness training can bring about long-term change in the brain (especially with regards to neuroplasticity) are being published.[39] Structural changes due to long term meditation (defined as more than 10,000 h of meditation) as well as functional changes and (more recently) changes in connectivity between different brain areas after much shorter training sessions have been described. The integration of neuroimaging studies with first person reports of what is experienced during meditation (as well as with behavioural changes) is on the frontline of the new research field called *contemplative science*.[40]

---

[38]For a discussion on the complexity of risk assessments and appropriate steps and measures see e.g. Hansson, S. O. (2004). Philosophical perspectives on risk. *Techné: Research in Philosophy and Technology*, *8*(1), 10–35.

[39]See for example Brefczynski-Lewis, J. A., Lutz, A., Schaefer, H. S., Levinson, D. B., & Davidson, R. J. (2007). Neural correlates of attentional expertise in long-term meditation practitioners. *Proceedings of the national Academy of Sciences*, *104*(27), 11483–11488. Slagter, H. A., Davidson, R. J., and Lutz, A. (2011). Mental training as a tool in the neuroscientific study of brain and cognitive plasticity. *Front. Hum. Neurosci.* 5:17. Ricard, M., Lutz, A., & Davidson, R. J. (2014). Mind of the Meditator. *Scientific American*, *311*(5), 38–45.

[40]One description of contemplative sciences would be: Just as scientists make observations and conduct experiments with the aid of technology, contemplatives have long tested their own theories with the help of developed meditative skills of observation and experimentation. Contemplative science aims for a deep knowledge of mental phenomena, including a wide range of states of consciousness, and its emphasis on strict mental discipline counteracts the effects of conative (intention and desire), attentional, cognitive, and affective imbalances. The integration of (neuro)science and contemplation is often highlighted as the aim of contemplative sciences. Wallace, B. Alan. *Contemplative science: Where Buddhism and neuroscience converge*. Columbia University Press, 2007. Britton, W. B., Brown, A. C., Kaplan, C. T., Goldman, R. E., DeLuca, M.,

The proposition is that these physiological changes can affect, and possibly even alter, our cognitive capacities, e.g. increase cognitive flexibility, improve attention focus and heighten our ability (and propensity) to act compassionately. Indeed it is not implausible that such changes would impact our preferences—perhaps as a consequence of enabling us to make all things considered assessments of complicated situations and make us more sensitive to the relevant, if salient, features of a situation.[41] Were this to be the case we would, at least in some situations, be able to make better decisions without first having to spend a substantive amount of energy on trying to control our more base desires, i.e. less prone to act on impulse.

As will be elaborated on in Chap. 5 a number of such changes could be connected to an improved capacity for instilling and maintaining a range of character traits (primarily epistemic virtues) as identified by Aristotle and some modern virtue epistemologists. Examples include intellectual courage, intellectual honesty and impartiality. While it would be unlikely that such and other epistemic virtues would cancel out implicit moral bias and attitudes, they might enable us to deal with the situation in a better fashion (for a discussion on this see Chap. 5). For example, increased awareness of our inbuilt biological limitations in combination with a fuller virtue skillset would makes us more aware and thus able to organise society in a manner which, at the very least, does not amplify our weaknesses.[42]

Were we to be able to better control what Kahneman and Twersky call System 1 and, further, possibly overcome some of the effects of bias that plague both System 1 and System 2 thinking, we would plausibly be better off. We might, for example, be less vulnerable to weakness of will and suffer fewer bouts of what Aristotle called *akrasia* (the result of reason being dragged around by desire, for more on this please see Chap. 6).

1.5.5 Character Development

Evidently findings on the potential of meditation (and other techniques or methods facilitating cognitive flexibility) do not cancel out more traditional forms of habituation and it would presumably be wise to assume that acting morally will remain very challenging. Indeed, the building of what might be broadly conceived of as 'character' and a level of predictability, consistency and coherence in choices and behaviour most certainly requires a great deal discipline and stubborn effort. None the less these findings add an interesting dimension to some of Aristoteles key

(Footnote 40 continued)

Rojiani, R., … & Frank, T. (2013). Contemplative science: an insider prospectus. *New Directions for Teaching and Learning*, *2013*(134), 13-29.

[41]Both moral and (objectively) factual features presumably.

[42]See Chap. 6 for a discussion on how we might proactively shape society through various embedding structures which generate pro-social behavior.

ideas. Aristotle held that the good life is not just a set of actions—it is a set of actions performed by someone who does them because she correctly sees the point in doing them. Such insights might not come as a bolt in the night and, consequently, moral virtue is acquired in stages, through education and habituation, and has both cognitive and emotional dimensions. The life of virtue is a long-term commitment, the virtues need to be practised and fine-tuned over a lifetime. On the Aristotelian model the virtues are not unlike physical fitness in the sense that they need to be exercised every day.[43] In a similar, if more recent, vein Professor Robert Kegan at Harvard has developed an interesting model of adult development. The model describes several steps or transitions in adult development and, how for each step the number of subjects who are able to transform further is incrementally diminishing.[44] The practice of meditation or other forms of virtue enhancement might increase not only the number of subjects who would be able to transform, but also of the speed at which they would be able to this. By enhancing both meta-awareness and cognitive flexibility by regularly practising meditation, one might become able to view moral choices from new perspectives which, in turn could open the door to novel and possibly more creative solutions. A reduction of misplaced fear could be expected, thereby increasing the range of possible actions. Related to this, an increased body-awareness might enable a better and more alert mode when it comes to moral choices based on e.g. unfairness and disgust.

We propose that a kind of oscillation will occur.[45] By this we mean that there will be a repetitive variation between two different states, in this case between the cultivation of new capacities and the trying them out in real life. This to-ing and fro-ing continues—we gain more experiences which we reflect on and then try out, from the conclusions we gain new insights and so on. Presumably, this would be a lifelong developmental cycle (although the range of the oscillation might get more and more narrow) as character building is not a quick fix, nor can we prepare for all situations and resolve them in advance. This take fits well with some key Aristotelian views e.g. that the virtuous life is a lifelong commitment but it must also be recalled that doing the right thing is leading the good life. It is not a means to an end or some far off reward.

---

[43]This view is also highlighted by researchers who study the effect of meditation practices, although the "dose", e.g. the time spent meditating varies a lot, and is subject to continuous research. Vøllestad, J., Sivertsen, B., & Nielsen, G. H. (2011). Mindfulness-based stress reduction for patients with anxiety disorders: Evaluation in a randomized controlled trial. *Behaviour research and therapy*, *49*(4), 281–288. Boettcher, J., Åström, V., Påhlsson, D., Schenström, O., Andersson, G., & Carlbring, P. (2014). Internet-based mindfulness treatment for anxiety disorders: A randomized controlled trial. *Behavior therapy*, *45*(2), 241–253.

[44]Kegan, R., & Lahey, L. L. (2009). *Immunity to change: How to overcome it and unlock potential in yourself and your organization*. Harvard Business Press. Kegan, R (1994). *In over our heads: the mental demands of modern life*. Harvard University Press.

[45]We also want to add a developmental aspect, in that the oscillations in the different states have the potential to a continuing improvement and refinement of the warranted capacities.

In summary, what the scientific findings could indicate is that we stand a better chance to

(i) develop the type of stable dispositions that issue in action than what was previously thought, and
(ii) that doing the right thing in fact would become both easier and more pleasurable the more we practice.

In other words, that we might be able to actually change our ways in a stable, lasting and generalizable way as opposed to simply putting on a thin moral gloss which flakes off as soon as push comes to shove. In addition, this opens up for the possibility that many people through training and effort can affect their brains in a manner which plausibly is conducive both to the establishing and up-keeping of *new habits*. This could trigger a positive spiral or domino effect which then might spread throughout society (for more on the idea societal spin-off effects see Chap. 6).

But before further exploring the full potential of meditation as a means to cognitive enhancement some alternative methods must be introduced. Notably, even though we view meditation as the better option at the present time we certainly do not rule out that the various methods can be combined to good effect in the not too distant future.

1.6 Enhancement Methods

Not surprisingly, the potential for structural and functional change has stirred a big interest both in the academic community and civil society. Indeed, the concept of neuroplasticity is one of the most googled terms in neuroscience in the last years.[46] However, sheer potential for change (and possible improvement) aside the next big question is *how*. How can we achieve such changes and how can we balance effect and risk? Attempting to address the problem various ways of improving decision-making skills have been suggested. Some are of a traditional nature, e.g. encouraging an increased commitment to behaving morally or creating incentives by tweaking the political structure, while others, involving pharmaceutical drugs[47] and nanotech brain implants, might come across as more radical.[48]

---

[46]https://www.google.com/trends/explore#q=Neuroplasticity.

[47]Heinrichs, M. et al. (2009) Oxytocin, vasopressin, and human social behavior. *Front. Neuroendocrinol.* 30, 548–557. Macdonald, K. and Macdonald, T.M. (2010) The peptide that binds: a systematic review of oxytocin and its prosocial effects in humans. Harv. Rev. Psychiatry 18, 1–21. Earp, B. D., Wudarczyk, O. A., Sandberg, A., & Savulescu, J. (2013). If I could just stop loving you: Anti-love biotechnology and the ethics of a chemical breakup. *The American Journal of Bioethics*, *13*(11), 3–17.

[48]Indeed, some such technologies might be less futuristic than one might think. In the last decade alone there have been significant breakthroughs in so called BMI (brain machine interface) technologies. Lebedev, M. A., & Nicolelis, M. A. (2006). Brain–machine interfaces: past, present

Clearly we need to become 'better' (as in more rational and virtuous) decision-makers both for our own sake and that of society as a whole. First and foremost we need to get better at actually doing that which we (know) we ought to do. We need to foster a more pro-social behavior and improve our life-skills as broadly conceived of.[49] Notably, however, it is of importance to strike the right balance between, on the one hand, taking a step back and detaching from our short-term interests and temptations and, on the other, becoming unresponsive to contextual features and too cold in our assessment of the situation. What is the best strategy to enable us to achieve this goal?

Without ruling out the future potential of cognitive enhancement through pharmaceutical drugs and technology, our position is that the combination of the pressing nature of the problem and the stage of where the science is today our best bet for lasting cognitive enhancement will come through 'life-style' broadly conceived of.

That said, it should be noted that the focus on life-style does not imply that we believe that pharmaceutical drugs (e.g. modafinil), hormones, neurotransmitters, nootropics etc. cannot be effective for achieving cognitive improvements,[50] or that research and drug trials should be stopped. Further, we also note the increasing interest in, and development of, technical devices such as sensors and actuators enabling increased body awareness, and smartphone applications who facilitate

---

(Footnote 48 continued)

and future. *TRENDS in Neurosciences*, *29*(9), 536–546. Vidal, J. J. (1973). Toward direct brain-computer communication. *Annual review of Biophysics and Bioengineering*, *2*(1), 157–180. Clausen, J. 2010. Ethical brain stimulation- neuroethics of deep brain stimulation in research and clinical practice. European Journal of Neuroscience 32: 1152–1162. Bell, E., G. Mathieu, and E. Racine. 2009. Preparing the ethical future of deep brain stimulation. Surgical Neurology 72: 577–586. Glannon, W. 2009. Stimulating brains, altering minds. Journal of Medical Ethics 35: 289–292. Very generally speaking, these are techniques which enable researchers to connect machines to the human nervous system. Nair, P. (2013). Brain–machine interface. *Proceedings of the National Academy of Sciences*, *110*(46), 18343–18343. The machines are then used to stimulate the brain and already today BMIs are used to treat deafness, Parkinson's disease and depression. Berger F. et al. (2008) Ethical, Legal and Social Aspects of Brain-Implants Using Nano-Scale Materials and Techniques, *NanoEthics*. 2,3, 241–249. E.C. Leuthardt et al., "Using the electrocorticographic speech network to control a brain-computer interface in humans," *J Neural Eng*, 8:036004, 2011. Collinger, J. L., Kryger, M. A., Barbara, R., Betler, T., Bowsher, K., Brown, E. H., … & Boninger, M. L. (2014). Collaborative Approach in the Development of High-Performance Brain–Computer Interfaces for a Neuroprosthetic Arm: Translation from Animal Models to Human Control. *Clinical and translational science*, *7*(1), 52–59. Iuculano, T., & Kadosh, R. C. (2013). The mental cost of cognitive enhancement. *The Journal of Neuroscience*, *33* (10), 4482–4486. Kadosh, R. C., Levy, N., O'Shea, J., Shea, N., & Savulescu, J. (2012). The neuroethics of non-invasive brain stimulation. *Current Biology*, *22*(4), R108-R111.

[49]As discussed in Chaps. 5 and 6 the instilling of the epistemic virtues will help to avoid falling into the trap of 'misplaced loyalty' in the broad sense.

[50]At present two of the major challenges are (a) that the effects in the normally functioning brain i.e. non therapeutic use, of the pharmaceutical drugs tend to have a comparatively small effect and (b) that the pharmaceutical drugs only affect single cognitive capacities/cognitive functions i.e. not overall intelligence. For more on this, please see Chap. 3.

(and remind the user of) mental training exercises, thereby supporting the life style changes we advocate. In Chap. 3 we will take a closer look at the competition, give concrete examples and comment further on the possibility of combinations.

1.6.1 A Few Words on Life Style Choices

Very broadly speaking life-style changes tailored to enhance our cognitive capacities can be split into (i) diet[51] and supplements[52] and (ii) the rest ranging from cardio-vascular exercising to memory training.[53] In this book we will not look at any diet or supplement strategies nor on physical exercise in general. We are interested in exploring the beneficial effects of meditation and, more specifically, of the three following meditation techniques; Attention Training Meditation, Open Monitoring and Compassion Training. The hypothesis is that engaging in regular meditation (using these techniques) can result in improved cognitive flexibility, which in turn, could be highly conducive to the instilling and upholding of epistemic traits such as introspection, intellectual courage, intellectual honesty and impartiality.

Evidently these are not the only methods with proven effects and for comparison we will also provide a brief account of another lifestyle change/habit which has generated positive results (in Chap. 3). The example looks at the connection between playing computer games (of a special kind) and improved cognition (especially memory training and focus attention). There is evidence of positive effects on e.g. process/working memory, problem solving, focus attention and creativity.[54] We find this example interesting for several reasons. Firstly there is a heated and long-standing debate on whether or not young people should be restricted in computer time as they sacrifice other (implicitly more valuable) activities, that playing games stumps emotional development and increase

[51]Positive effects which are more short-term are identified in studies focusing on the relationship between diet, brain capacity (e.g. attention- and process-memory) and performance. For links between cognitive capacity and glucose, creatine and amino acids (for example) see Fox, P.T., Raichle, M.E. et al. (1988). Nonoxidative glucose consumption during focal physiologic neural activity. Science, 241(4864), 462–4; McMorris, T., Harris, R.C. et al. (2006). Effect of creatine supplementation and sleep deprivation, with mild exercise, on cognitive and psychomotor performance, mood state, and plasma concentrations of catecholamines and cortisol. *Psychopharmacology*, 185(1), 93–103.

[52]Luchtman, D. W., & Song, C. (2013). Cognitive enhancement by omega-3 fatty acids from child-hood to old age: findings from animal and clinical studies. *Neuropharmacology*, *64*, 550–565.

[53]Tolppanen, A. M., Solomon, A., Kulmala, J., Kåreholt, I., Ngandu, T., Rusanen, M., … & Kivipelto, M. (2014). Leisure-time physical activity from mid-to late life, body mass index, and risk of dementia. *Alzheimer's & Dementia*.

[54]Granic, I., Lobel, A., & Engels, R. C. (2014). The benefits of playing video games. *American Psychologist*, *69*(1), 66.

anti-social behavior and so on.[55] Secondly, the amount of people who play computer games a regular basis is growing fast and most age groups are represented.

That said, these activities do not have to be mutually exclusive and there is no apparent reason to assume that they 'cannibalize' on each other. There is no inherent conflict in playing computer games *and* doing meditation. In other words, time and socio-economic circumstances permitting, a prudent agent could of course engage in all of them and others to for that matter.[56]

1.6.2 To What Effect

Essentially then, what we suggest is that most people could acquire a set of 'life-skills'—the ones mentioned here are just a few examples—which all things considered are likely to enable them to fare better in life. This could also be conducive to fostering pro-social behavior as broadly conceived of. An example would be how a better understanding of the underlying causes of climate change, coupled with an improved capacity to look beyond the self, would issue in a behavior which is more responsible. Further, given the uncertainty of what is to come it appears wise to acquire a broad skill-set as to maximize versatility, and these meditation techniques are highly efficient ways of doing this.

Evidently, some meditation techniques could have adverse effects and it is far from clear that they are suited for people who have underlying psychological or psychiatric issues, at least without professional guidance and structured follow up. A successful training in healthy people might also require an experienced teacher, or at least the possibility to consult such a person when needed. While we are aware of this aspect and consider it very important, we will, primarily for space reasons, only be able to touch briefly on such risks.

In the discussion part (Chap. 6) we briefly explore the role of social 'embedding structures' which could influence the development of these life-skills. Consider, for example, how a system that rewards generosity and self-lessness and offers special education packages on moral development could be conducive to the cultivation of good life-skills. Consider also the potential for the reverse effects in societies where,

---

[55]Spence, E.H. (2012), "Virtual rape, real dignity: meta-ethics for virtual worlds", The Philosophy of Computer Games, Springer, Dordrecht, pp. 125–142. Sicart, M. (2009), The Ethics of Computer Games, MIT Press. De Decker, E., De Craemer, M., De Bourdeaudhuij, I., Wijndaele, K., Duvinage, K., Koletzko, B. and Cardon, G. (2012), "Influencing factors of screen time in preschool children: an exploration of parents' perceptions through focus groups in six European countries", Obesity Reviews, Vol. 13, pp. 75–84.

[56]For a longer discussion on the potential for combinations of the various methods please see Fröding, B. (2013). *Virtue ethics and human enhancement*. Springer. Especially in chapter 7.

for example, free-riding, intimidation and greed is, if not encouraged, at least rewarded and not properly prohibited.[57]

1.6.2.1 Additional Reasons to Opt for Meditation

Somewhat mundane, but none the less important, reasons for why it might be an attractive strategy to opt for meditation include;

I. It is actually available (both in the practical/easy-to-learn sense and affordable)[58] and (comparatively) efficient. But, one could counter, so are pharmaceutical drugs like modafinil and methylphenidate (Ritalin, Concerta etc.), which have an effect, although not substantial, on the normal brain. But our point is rather this: the pharmaceutical drugs we have today might enhance (temporarily) one specific function e.g. awakeness or focus or memory—so while it might increase effectiveness it has no significant effect on the overall intelligence of the healthy individual. What meditation techniques potentially could create is quite different to that—namely a generalizable capacity[59] to make better all things considered choices. This capacity would, once the epistemic virtues are instilled, extend to various domains in life and also provide a strong incentive to keep on meditating (for more on the potential loop effects see Chap. 6).
II. It is low risk (both with regards to negative side effects and opportunity costs).[60] Again, it could be pointed out, so is modafinil. At the time of writing this we are not aware of any study that conclusively show that regular users become addicted or that they need to up the dose to maintain the effect. That said, it must immediately be added that the cohorts are very small and that they do not tell us much about the long-term effects on a larger population. There has also been reports on manic episodes etc. and it appears wise to be cautious at this stage.[61]

---

[57]See e.g. Zimbardo P. (2007) *The Lucifer Effect: Understanding How Good People Turn Evil.* Random House.

[58]Price is a big concern when it comes to pharmaceutical drugs and technology. The worry is that it would create even greater imbalance between the haves and the have not's. The counter argument is that everyone would be better off if some people (in decision-making positions presumably) would improve their cognitive capacities.

[59]Note that the capacity we are primarily interested in is cognitive flexibility and how that can be conducive to epistemic virtue. In other words we are not looking at a general increase of IQ and what may be the effects of such changes. However, we have identified one study where relational frame training (which is the theoretical basis for Acceptance and commitment therapy) seem to increase the IQ of the participating students. Cassidy, S., Roche, B. & Hayes, S. C. (2011). A relational frame training intervention to raise Intelligence Quotients: A pilot study. *The Psychological Record*, 61, 173–198.

[60]Barring people with mental issues as pointed out above in Sect. 1.5.2.

[61]Bell, S. K., Lucke, J. C. and Hall, W. D. (2012). Lessons for enhancement from the history of cocaine and amphetamine use. *AJOB Neuroscience*, 3(2): 24–29. Mohamed, A. D., & Sahakian, B. J. (2012). The ethics of elective psychopharmacology. *The International Journal of*

III. Observing recent progress and the ever growing interest in medicine and technology that can enable us to perform better it appears that we have good reasons to suspect that we will face some hard choices in the not too distant future.[62] While not imminent, pharmaceutical drugs and technology options will mature and become both available and much more potent (here we are thinking both about improvement of intelligence and overall computing capacity). Engaging in meditation and compassion training of the kind exemplified here would, at the very least, presumably not leave us worse off when we face this situation. Quite to the contrary, it appears that with regards to the capacity to make rational, all things considered choices regarding which pharmaceutical drugs and technology we ought to use, for what purposes, to what extent, how to assess risk and so on were we to commit to the life-style changes suggested here, would increase.

1.7 Which Cognitive Capacities?

Very generally speaking we will explore the effects that life-style changes (in the specific form of regular meditation) could have on cognition (and thereby on behavior). For concreteness, this text will look closer at three mental training techniques:

- the capacity to focus your attention, (*meditation involving focused attention training*)
- the capacity for introspection and detaching from yourself, (*open monitoring, which is integrated in e.g. mindfulness training*)
- the capacity to act more compassionate in the broad sense (to others and oneself) (*compassion training*)

We propose that engaging in this type of training on a regular basis has a positive impact on so called 'core cognitive capacities'. Our examples include; cognitive flexibility, focus/attention (which we view as related to meta-awareness), 'controlled' mind-wandering[63] and emotional regulation.

(Footnote 61 continued)

Neuropsychopharmacology, 15*(04), 559–571. Scoriels, L., Barnett, J. H., Murray, G. K., Cherukuru, S., Fielding, M., Cheng, F., ... & Jones, P. B. (2011). Effects of modafinil on emotional processing in first episode psychosis.* Biological psychiatry, 69*(5), 457–464.*

[62] For some problems attaching to the Precautionary Principle see Sandin, P., Peterson, M., Hansson, S. O., Rudén, C., & Juthe, A. (2002). Five charges against the precautionary principle. *Journal of Risk Research*, 5(4), 287–299.

[63] Franklin, M. S., Mrazek, M. D., Anderson, C. L., Smallwood, J., Kingstone, A., & Schooler, J. W. (2013). The silver lining of a mind in the clouds: interesting musings are associated with positive mood while mind-wandering. *Frontiers in psychology, 4.*

These capacities will be defined and a brief account of why they might be useful to have to a larger extent will be provided. In addition, there will be concrete examples of how these improvements are both lasting and possible to generalize outside the laboratory situation.[64]

Reference

The World Economic Forum's *Global Risks 2013* Report http://www3.weforum.org/docs/WEF_GlobalRisks_Report_2013.pdf

[64]Flook, Lisa; Goldberg, Simon B.; Pinger, Laura; Davidson, Richard J. Promoting Prosocial Behavior and Self-Regulatory Skills in Preschool Children Through a Mindfulness-Based Kindness Curriculum. Developmental Psychology, Nov 10, 2014.

Chapter 2
The Neurophysiological Background

Abstract This chapter begins by providing a schematic review of the neurophysiological background for the development of a set of capacities e.g. cognitive flexibility, meta awareness and emotional regulation. This includes a brief account of the evidence based scientific studies showing that we *can* change—i.e. that the adult human brain is plastic enough. We have our biological set up, and it is probably a pre-requisite that we are getting to know our limitations as well as our possibilities, in order to perform the changes that are warranted. Then we turn to the question of *how* to bring about such changes. In this chapter we examine the changes that regular mental training in the form of meditation can have on the adult brain and behavior.

Keywords Neuroplasticity · Attention · Mind-wandering · Meditation · Compassion

2.1 Introduction

The combination of the increasing flow of information in society, on the one hand, and our relatively meagre cognitive capacities and emotional repertoire, on the other, effectively stops us from living as good and happy life as we could. Our collective failure to process the information and then make responsible decisions based on our conclusions[1] has disastrous consequences. We harm ourselves, each other and the planet. Although there are several very advanced examples of how

[1]Sheena, I. (2011). The art of choosing.

B. Fröding and W. Osika, *Neuroenhancement: How Mental Training and Meditation Can Promote Epistemic Virtue*, SpringerBriefs in Ethics,
DOI 10.1007/978-3-319-23517-2_2

e.g. "big data" could be handled in efficient ways,[2] in order to facilitate decision-making, it seems as we have a long way to go before such methods could be used seamlessly in everyday life. Evidently, it is not enough to make the right decision on occasion, but also to be able to persist in choosing the necessary new habits which might be connected to ones "valued direction".[3] At the same time one has to adapt to an ever changing environment (as well as ultimately be a co-creator of this environment).

As made clear in Chap. 1 humans would have much to gain from improved decision-making. In other words—the *why* question is answered. To become more skilled in this domain—for example less biased, more able to assess risk and better at epistemic deference—would be beneficial both for the individual and the collective. Concrete examples of cognitive capacities, which plausibly could contribute to responsible decision-making are cognitive flexibility, meta awareness and emotional regulation. These and other candidates will be further explored in the next chapters.

But before we can turn to the *how* (i.e. the main question of this book) we need to provide a schematic introduction of the neurophysiological background for the development of such cognitive capacities. We start with a brief account of the evidence based scientific studies showing that we *can* change (i.e. that the adult human brain is plastic enough) and then we turn to look at some techniques for prompting such change.[4]

2.2 Neuroplasticity

The adult human brain is affected by environmental changes and pressures, physiologic modifications and experiences, which lead to functional and structural reorganization, i.e. *neuroplasticity.* The paradigm of neuroplasticity is

[2]Manyika, J., Chui, M., Brown, B., Bughin, J., Dobbs, R., Roxburgh, C., ... & McKinsey Global Institute. (2011). Big data: The next frontier for innovation, competition, and productivity. Lampitt, A. (2013). 'The real story of how Big Data analytics helped Obama win'. *Think big data-infoworld.* Himelfarb S. (2014) Can big data stop wars before they happen? *Foreign Policy.* One of the first and most well-known applications for big data is the flu spotting algorithms developed by Google together with The Centers for Disease Control and Prevention. Ginsberg, J., Mohebbi, M. H., Patel, R. S., Brammer, L., Smolinski, M. S., & Brilliant, L. (2009). Detecting influenza epidemics using search engine query data. *Nature*, *457*(7232), 1012–1014. Also in neuroscience extremely large amounts of data from e.g. neuroimaging-behavioral experiments are derived and analyzed in relation to functionality vs locality and connectivity.

[3]With "valued direction" we here mean the individuals valued direction in life. See also Chap. 1.

[4]Evidently the argument that some changes in the brain (as described here) can—to an extent—be connected to improvement in behavior rests on some assumptions on the possibility of free will and moral responsibility. The larger philosophical discussion on this falls well outside the scope of this volume but for a very interesting account of an agency cultivation model (of how holding someone morally accountable for their actions and the effects of practices matters when it comes to cultivating moral agency) see Vargas, M. (2013). *Building better beings: A theory of moral responsibility*. Oxford University Press.

fundamentally altering a long-held belief that changes in the central nervous system is only possible during critical periods during early development, although some forerunners discussed this possibility decades ago.[5, 6, 7] It is the mechanism for learning and for growth and development,[8] but also for effects of severe deprivation.[9]

Over the last decades novel neuroimaging techniques have been used to unravel experience dependent plasticity, and the number of studies is growing continuously. Hence, we have learnt a lot about the mechanisms of plasticity and the relations between brain activity and behaviour.

It seems that formation of new pathways is possible only following initial reinforcement of pre-existent connections. Possible plastic changes are limited by existing connections, which are the result of genetically controlled neural development and are ultimately different across individuals. Reinforcement of existing connections, on the other hand, is the consequence of environmental influences, such as education, training, observing others or an enrichment of the environment.[10]

There are different modes in how this reinforcement influences the shape and function of the brain. One example from animal studies is how the outside world impacts on rats, which dates back to the 1940s, when Donald Hebb showed that free range rats (compared with the ones in standard lab cage) after some weeks in

---

[5]Already in 1783 Bonnet and Vicenzo discovered that dogs and birds that had been trained had an increase of cerebellar "folds", a finding that at that time did not change the idea that the brain was unchangeable. Doidge, N. (2007). *The brain that changes itself: Stories of personal triumph from the frontiers of brain science*. Penguin.

[6]More than a decade later James (1890) in The Principles of Psychology introduced the term plasticity to the neurosciences in reference to the susceptibility of human behaviour to modification: Plasticity [. . .] means the possession of a structure weak enough to yield to an influence, but strong enough not to yield all at once. Each relatively stable phase of equilibrium in such a structure is marked by what we may call a new set of habits. Organic matter, especially nervous tissue, seems endowed with a very extraordinary degree of plasticity of this sort; so that we may without hesitation lay down as our first proposition the following, that the phenomena of habit in living beings are due to the plasticity (William James, 1890, The Principles of Psychology, Habit, Chap. 4, p. 68).

[7]Kheirbek, M. A., & Hen, R. (2013). (Radio) active neurogenesis in the human hippocampus. *Cell, 153*(6), 1183–1184.

[8]Pascual-Leone, A., Amedi, A., Fregni, F., & Merabet, L. B. (2005). The plastic human brain cortex. *Annu. Rev. Neurosci., 28*, 377–401.

[9]Mehta, M. A., Golembo, N. I., Nosarti, C., Colvert, E., Mota, A., Williams, S. C., … & Sonuga Barke, E. J. (2009). Amygdala, hippocampal and corpus callosum size following severe early institutional deprivation: the English and Romanian Adoptees study pilot. *Journal of Child Psychology and Psychiatry, 50*(8), 943–951.

[10]There is an emerging field of connectivity studies, where complex networks are being conceptualized, parallel to the development of both measurement methods and algorithms for analysis of the huge amounts of data these studies are producing. Sporns, O. (2013). The human connectome: origins and challenges. *Neuroimage, 80*, 53–61.

his home had superior problem solving abilities, such as maze running.[11] The term "environmental enrichment" was later used to describe this phenomenon, which seems to be of special importance and powerful when the environment is stimulating, novel and spurs exploration. It seems also that it is of importance that the exercise is voluntary: to generate neuronal changes the animal must decide to enter e.g. the exercise wheel and run in it.[12] In humans it is harder to perform ethically sound studies where only one group will receive the stimulation of enriched environments. However, increasing evidence from observational studies point to that greater involvement in intellectual and social activities is linked to a slower cognitive decline, in elderly populations,[13] but also aerobic exercises has been shown to be of benefit for brain volume in aging humans.[14]

Maguire et al. demonstrated that experience shapes the human brain in a study of London taxi drivers. The drivers' detailed knowledge of London's street plan was reflected in enlarged posterior hippocampi (an area of importance for memory storage) and the size of the hippocampi correlated with the number of years spent driving taxi.[15] In a longitudinal study investigating juggling training over a three-month period, an increase of grey matter in the visual motion areas was shown.[16] Wollet and Maguire again studied taxi drivers, but this time they captured data at the beginning and at the end of the drivers 3–4 years training. Increased posterior hippocampi were only observed in trainees who qualified the tests, but not in trainees who failed.[17]

Musical training induces cortical plasticity and increases functional connectivity between areas in the brain.[18] The learning of more abstract information has also

---

[11]Pickren, W., & Rutherford, A: (2010). A history of modern psychology in context. Hoboken, NJ: Wiley.

[12]Van Praag, H., Kempermann, G., & Gage, F.H. (1999). Running increases cell proliferation and neurogenesis in the adult mouse dentate gyrus. *Nature Neuroscience,* 2(3), 266–270.

[13]Nyberg, L., Lövdèn, M., Riklund, K., Lindenberger, U., & Bäckman, L. (2012). Memory aging and brain maintenance. Trends in Cognitive Sciences. 16(5), 292–305.

[14]Colcombe, S. J., Erickson, K. I., Scalf, P. E., Kim, J. S., Prakash, R., McAuley, E., … & Kramer, A. F. (2006). Aerobic exercise training increases brain volume in aging humans. *The Journals of Gerontology Series A: Biological Sciences and Medical Sciences*, *61*(11), 1166–1170.

[15]Maguire, E. A., Gadian, D. G., Johnsrude, I. S., Good, C. D., Ashburner, J., Frackowiak, R. S., & Frith, C. D. (2000). Navigation-related structural change in the hippocampi of taxi drivers. *Proceedings of the National Academy of Sciences*, *97*(8), 4398–4403.

[16]Draganski, B., Gaser, C., Busch, V., Schuierer, G., Bogdahn, U., & May, A. (2004). Neuroplasticity: changes in grey matter induced by training. *Nature*, *427*(6972), 311–312.

[17]Woollett, K., & Maguire, E. A. (2011). Acquiring "the Knowledge" of London's layout drives structural brain changes. *Current biology*, *21*(24), 2109–2114.

[18]Lappe, C., Herholz, S. C., Trainor, L. J., & Pantev, C. (2008). Cortical plasticity induced by short-term unimodal and multimodal musical training. *The Journal of Neuroscience*, *28*(39), 9632–9639. Pinho, A. L., de Manzano, Ö., Fransson, P., Eriksson, H., & Ullén, F. (2014). Connecting to Create: Expertise in Musical Improvisation Is Associated with Increased Functional Connectivity between Premotor and Prefrontal Areas. *The Journal of Neuroscience*, *34*(18), 6156–6163.

been shown to be associated in neuroplasticity. After intensive studying for an exam, increases in grey matter volume was shown in several regions in medical students, and three months of language learning resulted in increased cortical thickness in frontal and temporal regions, as well as an increase in hippocampal volumes.[19] However, there are also negative studies that did not show any structural changes in grey matter after training. One explanation of these negative results is related to the type of interventions.

Mental simulation of movements activates some of the same central neural structures required for the performance of the actual movements.[20] In doing so, mental practice alone may be sufficient to promote the plastic modulation of neural circuits placing the subjects at an advantage for faster skill learning with minimal physical practice, presumably by making the reinforcement of existing connections easier and perhaps speeding up the process of subsequent sprouting and consolidating of memories. Pascual Leone in a seminal experiment with piano playing, showed that *mental* practice resulted in a similar reorganization of the cortical motor outputs to the one observed in the group of subjects that physically practiced the movements.[21]

Recent neuroimaging studies have shown that mental images are accompanied by processes which, in some respects, are similar to those involved in the initial perception of *sensory* events, and in recalling those images from autobiographical memory.[22] One example is that visual mental imagery can activate areas in early visual cortex e.g. when making comparative judgments of imagined shape.[23]

2.3 Attention

The capacity to focus one's attention was defined by William James as "the sudden taking possession by the mind, in clear and vivid form, of one of what seems simultaneously possible objects or trains of thought".

Goleman (2013) describes the two main varieties of distractions of our attention: sensory and emotional, where the sensory can be seen as "an endless wave of

---

[19]Mårtensson, J., Eriksson, J., Bodammer, N. C., Lindgren, M., Johansson, M., Nyberg, L., & Lövdén, M. (2012). Growth of language-related brain areas after foreign language learning. *Neuroimage*, *63*(1), 240–244.

[20]Decety J, Ingvar DH. 1990. Brain structures participating in mental simulation of motor behavior: a neuropsychological interpretation. Acta Psychol. (Amst.) 73:13–34.

[21]Pascual-Leone, A., Nguyet, D., Cohen, L. G., Brasil-Neto, J. P., Cammarota, A., & Hallett, M. (1995). Modulation of muscle responses evoked by transcranial magnetic stimulation during the acquisition of new fine motor skills. *Journal of neurophysiology*, *74*(3), 1037–1045.

[22]Holmes, E. A., & Mathews, A. (2010). Mental imagery in emotion and emotional disorders. *Clinical psychology review*, *30*(3), 349–362.

[23]Kosslyn, S. M., Ganis, G., & Thompson, W. L. (2001). Neural foundations of imagery. *Nature Reviews Neuroscience*, *2*(9), 635–642.

incoming stimuli your brain weeds out from the continuous wash of background sounds, shapes and colours, tastes, smells and sensations etc.". He describes the emotional loaded signals as a much more daunting, and harder to divert from, especially if it comes from close relationship turmoil. This signalling is also an example of what neuroscientists call "bottom-up" processes—information from the peripheral nervous system processed by lower brain areas, and often never reaching the level of conscious experience (some researchers do also include e.g. pharmaceutical products affecting the body unconsciously, although expectations when ingesting a medicine is known to be very powerful, and demonstrated in many experiments on placebo).

The opposite would be top-down processes, a voluntarily act of task/goal directed behaviour, and the quote above from James (1890) also indicates that there is a voluntary part of the attentional process.

It is also interesting to note that the brain cannot distinguish real stimuli or threats from the one we only think of.[24] This results in that our inner thought processes will feed fruitless ruminating processes in obsessive loops, unless we come up with tentative solutions and then can let those distressing thoughts go. The ability to inhibit emotions and stay on target is connected to better performance, and operates in the prefrontal lobes.[25]

The ability to prevent distraction depends on the current level of attentional control activity in frontal cortex,[26] and also on a suppressive mechanism that reduces the salience of potentially distracting factors. This has popularly been called an "anti-distraction mode" in our brain, and means that focusing on a chosen object is not only about intentionally paying attention to it, but also suppressing as many distractions in the background as possible.

2.4 Mind-Wandering

Mind-wandering (sometimes referred to as task-unrelated thought) is described as the experience of thoughts not remaining on a single topic for a long period of time, and tends to occur during driving, reading and other activities where vigilance may

---

[24]Gilbert, P., Baldwin, M. W., Irons, C., Baccus, J. R., & Palmer, M. (2006). Self-Criticism and Self-Warmth. An Imagery Study Exploring their Relation to Depression. *Journal of Cognitive Psychotherapy.*

[25]Begley, S., & Davidson, R. (2012). *The emotional life of your brain.* Hachette UK.

[26]Gaspar, J. M., & McDonald, J. J. (2014). Suppression of salient objects prevents distraction in visual search. *The Journal of Neuroscience, 34*(16), 5658–5666.

be low. It is connected to what William James called the train of thought and the stream of consciousness.[27]

The phenomenon of when the mind starts "wandering" is related to some negative outcomes, such as worsening in comprehension when students are tested.[28] At a first glance also our emotions seem to be negative during mind-wandering, according to the "Track your happiness" study, where several thousands of volunteers answered a plethora of question delivered at random times at their smartphones.[29] Ruminating over the past while mind-wandering is linked to negative mood. When studying mind-wandering during functional imaging (fMRI) the behavioural activity of the mind is mirrored by an activity in the "default mode network[30]" engaging the medial prefrontal cortex, but when being in an executive mode, other areas are activated (e.g. anterior cingulate cortex and dorsolateral prefrontal cortex).

However, it seems as if also this kind of almost "in-attention" can be very useful. Although the prefrontal cortex may be used during this mind-wandering in a, for our awareness unfocused way, it sometimes results in a surprising problem solving outcome. Especially if one has an interesting musing, that is related to a positive emotion during mind-wandering.[31]

---

[27]"Consciousness, then, does not appear to itself chopped up in bits. Such words as 'chain' or 'train' do not describe it fitly as it presents itself in the first instance. It is nothing jointed; it flows. A 'river' or a 'stream' are the metaphors by which it is most naturally described. *In talking of it hereafter let us call it the stream of thought, of consciousness, or of subjective life.*" James, William (1890), *The Principles of Psychology*. p. 239.

[28]Smallwood, J., Fishman, D. J., & Schooler, J. W. (2007). Counting the cost of an absent mind: Mind wandering as an underrecognized influence on educational performance. *Psychonomic Bulletin & Review*, *14*(2), 230–236.

[29]Killingsworth and Gilbert, (2010) A Wandering Mind Is an Unhappy Mind. *Science.* Vol. 330 no. 6006 p. 932.

[30]"The DMN is a large-scale brain network defined by the temporal correlation between two core regions on the medial surface of the cortex, known as the posterior cingulate and medial prefrontal cortex. These regions form the core of the DMN and interact with subnetworks. Meta-analyses have shown that the core of this system tends to be engaged in self-referential processes, the medial temporal subsystem is engaged by episodic processes, and the dorsal medial subsystem is engaged by social processes. Together, these forms of thought are similar to the content of the self-generated thoughts that often occur during mind wandering, providing important evidence for the involvement of these regions in the mental content that occurs during mind wandering." Smallwood, J., & Schooler, J. W. (2015). The science of mind wandering: empirically navigating the stream of consciousness. *Annual review of psychology*, *66*, 487–518.

[31]Franklin, M. S., Mrazek, M. D., Anderson, C. L., Smallwood, J., Kingstone, A., & Schooler, J. W. (2013). The silver lining of a mind in the clouds: interesting musings are associated with positive mood while mind-wandering. *Frontiers in psychology, 4.*

2.5 Emotional Regulation

Attention is also of major importance for emotional self-regulation,[32] and this is underlined in several scientific disciplines, where emotional regulation is described as critical to general wellbeing.[33, 34]

If one is able to notice one owns emotion, and also to some extent influence the emotional self-regulating process, that can potentially facilitate decision making processes.[35] This regulation habit can however also go astray. We as a species characteristically make efforts to control emotion, and if we experience emotions being too unpleasant, we will use strategies in order to avoid them, a behaviour which has been described as "experiential avoidance". Avoiding the experience of tough and anxiety-related thoughts and feelings, and the associated diminished range of choices and behaviours, has in turn been shown to be associated to a significant number of problems, such as increased social withdrawal and impaired ability to handle stress.[36] So, the ability to both identifying emotions and notice the tendency to avoid the unpleasant ones provides a useful base for a successful emotional regulation.

Emotional regulation starts early. For an everyday example consider how infants already at 3–6 months of age can be soothed from distress induced by the presentation of novel objects.[37] Such external cues to re-orienting are one of the ways parents/caregivers support their child to develop self-regulation. This also has a neural basis in the network that is involved in orienting to sensory events, which is accompanied by the alerting network, involved in achieving and maintaining the alert state.

---

[32]Frank, D. W., Dewitt, M., Hudgens-Haney, M., Schaeffer, D. J., Ball, B. H., Schwarz, N. F., … & Sabatinelli, D. (2014). Emotion regulation: Quantitative meta-analysis of functional activation and deactivation. *Neuroscience & Biobehavioral Reviews*, *45*, 202–211.

[33]Bandura, A., Caprara, G. V., Barbaranelli, C., Gerbino, M., Pastorelli, C. (2003). "Role of Affective Self-Regulatory Efficacy in Diverse Spheres of Psychosocial Functioning". *Child Development* **74** (3): 769–82.

[34]Emotions are defined as broadly integrative systems ordering feeling, thought and action, and representing the output of information processing assessing the meaning or affective significance of events for the person. LeDoux, J. E. (1989). Cognitive-emotional interactions in the brain. *Cognition & Emotion*, *3*(4), 267–289.

[35]Gross, J. J. (2002). "Emotion regulation: Affective, cognitive, and social consequences". *Psychophysiology* 39 (3): 281–91. Miclea, M, Miu, A. (2010). "Emotion Regulation and Decision Making Under Risk and Uncertainty". *Emotion* 10 (2): 257–65.

[36]Hayes, Steven C.; Wilson, Kelly G.; Gifford, Elizabeth V.; Follette, Victoria M.; Strosahl, Kirk (1996). "Experiential avoidance and behavioral disorders: A functional dimensional approach to diagnosis and treatment". *Journal of Consulting and Clinical Psychology* 64 (6): 1152–68. See also Chap. 1, p 4, for a description of experiential avoidance.

[37]Harman, C., Rothbart, M. K., & Posner, M. I. (1997). Distress and attention interactions in early infancy. *Motivation and Emotion*, *21*(1), 27–44.

A shift in control from the brain's orienting network to the executive network (which functions to monitor and resolve conflicts between other brain networks) happens when the child is between 3 and 4 years old.[38] The shift contains the evolving abilities to ignore distractions, inhibit responses and impulses, and the ability to plan. With increasing self-awareness, both metacognitive (thinking about thinking) and meta-emotion (noticing, and understanding the stream of feeling and impulses[39]) skills can be developed. Those together can be described as self-management, where executive attention enables us to focus on to one thing and ignore others.[40] The strategic allocation of attention has also been shown to be of importance when testing self-restraint against instant gratification. In a series of seminal studies referred to as the "Marshmallow Test", Mischel et al.[41] invited four-year-olds to study their ability to resist the lure of a marshmallow, and could show that the ones who could delay the gratification of eating a marshmallow, as a reward got an extra marshmallow after 15 min had higher scores on measures of executive control, especially the reallocation of attention. At follow up, during several decades, it has been elucidated that these abilities link to development over the life course and predict many outcomes (e.g. SAT scores, social and cognitive competence, educational attainment and drug use) (Mischel 2014). In their forties, there were also differences in neuronal processing when impulse control was tested during fMRI studies, with a greater involvement of higher cortical areas in the "delayer".

So, what happens in the human brain when we are able to delay gratification? What seems to be needed is an ability to voluntarily disengage from the object which should be "avoided", then let our focus prevail elsewhere and resist to gravitate back to the desired object, and remember the reward or goal that awaits in the future.[42] Some amount of self-awareness and meta-skills seem also to facilitate this process.

---

[38]Rothbart, M. K., Sheese, B. E., Rueda, M. R., & Posner, M. I. (2011). Developing Mechanisms of Self-Regulation in Early Life. *Emotion Review*, *3*(2), 207–213.

[39]The opposite would be *alexithymia*, the inability to identify and describe emotions in the self. FeldmanHall, O., Dalgleish, T., & Mobbs, D. (2013). Alexithymia decreases altruism in real social decisions. *Cortex*, *49*(3), 899–904.

[40]Executive control varies between individuals, and a heritability has been shown. Fan, J., Wu, Y., Fossella, J. A., & Posner, M. I. (2001). Assessing the heritability of attentional networks. *BMC neuroscience*, *2*(1), 14.

[41]Mischel, W., Shoda, Y., & Rodriguez, M. L. (1989). Delay of gratification in children. *Science*, 244, 933–938. In the follow up study 59 subjects underwent a behavioral study, and a brain imaging study on a sub-group from the behavioral study (N = 27) was performed: Casey, B. J., Somerville, L. H., Gotlib, I. H., Ayduk, O., Franklin, N. T., Askren, M. K., ... & Shoda, Y. (2011). Behavioral and neural correlates of delay of gratification 40 years later. *Proceedings of the National Academy of Sciences*, *108*(36), 14998–15003.

[42]See also section on Focused Attention Meditation below.

2.6 Mental Training—Meditation

At long last, it is time to examine the changes that regular mental training in the form of e.g. meditation can have on the adult brain and behavior. In nearly all major religions and also in several philosophical traditions, different forms of meditation have been a core part of their practice (Zajonc 2009).

One way of defining meditation is the cultivation of basic human qualities, such as a more stable and clear mind, emotional balance, cognitive flexibility and a sense of compassion—qualities that can remain more or less latent, but also could be developed by regular practice.[43]

We will discuss three examples of practices where meditation is a core component. In this process we have chosen to restrict our references regarding biological effects of the practices to the structural and functional neuroimaging studies, which now enhance our understanding of the neural processes associated with meditation.[44] We would also highlight that "traditional Buddhist formulations describe meditation as a state of relaxed alertness that must guard against both excessive hyperarousal (restlessness) and excessive hypoarousal (drowsiness, sleep). The impression is that modern applications of meditation have emphasized the hypoarousing and relaxing effects without as much emphasis on the arousing or alertness-promoting effects. However, there are findings that suggest that the course of meditative progress exhibits a nonlinear multiphasic trajectory, such that early phases that are more effortful may produce more fatigue and sleep propensity, while later stages produce greater wakefulness as a result of neuroplastic changes and more efficient processing" (Britton et al. 2014).

2.6.1 *Focused Attention Meditation*

The development of attention skills is one of the central components of many meditation traditions such as mindfulness meditation practice.[45] The aim of training of attention skills is e.g. to enhance the capability to sustain non-judgmental awareness of one's thinking patterns, emotions, and sensory perceptions, and to

---

[43]A common way to start meditating is by assuming a comfortable physical posture. A "self-priming" of the mind is often recommended, such as instilling a wish for an increased self-knowledge in order to develop the skills mentioned above. Stabelizing of the mind, which often is unfocused and disturbed by a never-ending inner chatter, is a major objective. Debarnot, U., Sperduti, M., Di Rienzo, F., & Guillot, A. (2014). Experts bodies, experts minds: how physical and mental training shape the brain. *Frontiers in human neuroscience*, *8*. Ricard, M., Lutz, A., & Davidson, R. J. (2014). Mind of the Meditator. *Scientific American*, *311*(5), 38–45.

[44]Marchand, W. R. (2014). Neural mechanisms of mindfulness and meditation: Evidence from neuroimaging studies. *World journal of radiology*, *6*(7), 471.

[45]Slagter, H. A., Davidson, R. J., and Lutz, A. (2011). Mental training as a tool in the neuroscientific study of brain and cognitive plasticity. *Front. Hum. Neurosci.* 5:17.

centre it in the present moment, while developing the capacity to be vigilant to both inner and external distractions.[46] Malinowski et al. have developed a model of how the training of attention skills is thought to underpin emotional and cognitive flexibility, bringing about the ability to maintain non-judging awareness of one's own thoughts, feelings, and experiences in more general terms. They also imply how performance increases described above are reflected as changes in neural activity and underlying neural architecture. Hasenkamp et al. describes four cognitive cycle intervals relevant for meditation: mind wandering, awareness of the wandering of one's mind, varying of attention, and prolonged attention.[47] As one practices the flexible shifting of attention, for instance, from attention to the breath to attention to sounds, cognitive flexibility is also increased (for more on cognitive flexibility see Chap. 4). When attaining sharp focus, key circuitry in the prefrontal lobes gets into a synchronized state, "phase-locking".[48]

This is especially interesting as metacognitive abilities develop during childhood and adolescence as the prefrontal cortex matures and executive functions develop. An increase of meta-skills due to meditation has been described, and it could be hypothesized that the number of possible metacognitive levels is related to e.g. the working memory capacity and effectiveness of the central executive.[49]

The study of performance during an fMRI adapted stress task, which requires impulse and attention control showed that meditation improves efficiency, perhaps by enhancing the ability to sustain attention and control impulses.[50] Also working memory functioning (which is impaired by e.g. stress) has been shown to be positively affected by meditation practices.[51]

---

[46]We are aware of that selecting just one single capacity, attention, of course will lead to a simplified description of a rather complex process, but we do it for pedagogical reasons. For example, as focused attention and open monitoring conceptually can be separated, even simple forms of mindfulness training will entail both components. Malinowski, P. (2013). Neural mechanisms of attentional control in mindfulness meditation. *Frontiers in neuroscience, 7.*

[47]Hasenkamp, W., Wilson-Mendenhall, C. D., Duncan, E., & Barsalou, L. W. (2012). Mind wandering and attention during focused meditation: a fine-grained temporal analysis of fluctuating cognitive states. *Neuroimage, 59*(1), 750–760.

[48]Slagter, H. A., Lutz, A., Greischar, L. L., Nieuwenhuis, S., & Davidson, R. J. (2009). Theta phase synchrony and conscious target perception: impact of intensive mental training. *Journal of Cognitive Neuroscience, 21*(8), 1536–1549.

[49]Brefczynski-Lewis, J. A., Lutz, A., Schaefer, H. S., Levinson, D. B., & Davidson, R. J. (2007). Neural correlates of attentional expertise in long-term meditation practitioners. *Proceedings of the national Academy of Sciences, 104*(27), 11483–11488. Jankowski T, Holas P. Metacognitive model of mindfulness. Conscious Cogn. 2014 Jul 16; 28C: 64–80.

[50]Kozasa, E. H., Sato, J. R., Lacerda, S. S., Barreiros, M. A., Radvany, J., Russell, T. A., … & Amaro, E. (2012). Meditation training increases brain efficiency in an attention task. *Neuroimage, 59*(1), 745–749.

[51]Working memory is the system that is responsible for the transient holding and processing of new and already stored information, an important process for reasoning, comprehension, learning and memory updating. Jha, A. P., Stanley, E. A., Kiyonaga, A., Wong, L., & Gelfand, L. (2010). Examining the protective effects of mindfulness training on working memory capacity and affective experience. *Emotion, 10*(1), 54.

2.6.2 Open Monitoring, E.G. in Mindfulness Training

Mindfulness has been described as a state of consciousness, experiencing the present moment, both external stimuli and a meta-awareness of one's internal thoughts and emotions without judging or trying to change the present. "The awareness that emerges through paying attention on purpose, in the present moment, and nonjudgmentally to the unfolding of experience moment by moment".[52] Mindfulness has its roots in eastern traditions and Buddhist practices but was adapted to a secular context free of religious components, and it may be developed during meditation and experienced during one's daily life.[53] The program *mindfulness based stress reduction* (MBSR) was developed by Jon Kabat-Zinn decades ago and was introduced in psychology and medicine as a coping resource for anxiety, stress and chronic pain.

Under stress, our capacity to see the perspective of others (theory of mind), to keep a sharp focus of the bigger picture (central coherence), and to be flexible and organize information and behaviour (executive functioning) is hampered.[54] In mindfulness training, participants become aware of signs of stress in their body, and by taking, for instance, a 3-min breathing space, a mini-meditation in which one gets out of automatic pilot and into the present moment,[55] stress may reduce and the perspective may be widened. In an ambitious study Catherine Kerr et al. studied the diaries of participants of a MBSR program, and discovered (after an initial resistance towards the experiences displayed by the training) the emergence of an observing mind, after about 5–6 weeks of practice.[56] In the eastern traditions meta-experiences has been described as the core component of the phenomenon called "the beginner's mind".[57]

---

[52]Kabat-Zinn, J. (2003). Mindfulness-based interventions in context: past, present, and future. *Clinical psychology: Science and practice*, *10*(2), 144–156.

[53]Williams, J. M. G. (2008). Mindfulness, depression and modes of mind. *Cognitive Therapy and Research*, *32*(6), 721–733.

[54]Brosschot J.F. (2010) Markers of chronic stress: Prolonged physiological activation and (un) conscious perseverative cognition. *Neuroscience and Biobehavioral Reviews* 35: 46–50. Liston C., McEwen B.S., Casey B.J. (2009) Psychosocial stress reversibly disrupts prefrontal processing and attentional control. *Proc Natl Acad Sci* USA 106: 912–917.

[55]Segal ZV, Williams JM, Teasdale JD (2012) *Mindfulness-based cognitive therapy for depression: a new approach to preventing relapse*. The Guilford Press.

[56]Kerr, C. E., Josyula, K., & Littenberg, R. (2011). Developing an observing attitude: an analysis of meditation diaries in an MBSR clinical trial. *Clinical Psychology & Psychotherapy*, *18*(1), 80–93.

[57]Suzuki, S. (2010). *Zen mind, beginner's mind*. Shambhala Publications.

2.6.3 Compassion

The capacity to notice suffering and act compassionate in the broad sense (to others and oneself).[58]

Emotional regulation, which is disrupted in most mental disorders, has been shown to improve after mindfulness training.[59] Some clinicians argue that it is wise to actually start meditation practice with compassion training,[60] in order to increase the functioning of the "self-soothing" system and thereby enable the meditator to face difficult experiences which might come with the increase of focused attention abilities.

Recent behavioural and neurophysiological research vindicates that having one's emotions resonate empathetically with the feelings of another person, and of compassion really differ. Compassion (and altruistic love) is associated with positive emotions, and to merely empathetically resonate with others (or one owns memories of) suffering can initiate processes of emotional exhaustion or burnout, in fact, a kind of empathy fatigue, or an emotional shut down.[61] An example of how such experiments are performed is a series of studies from Tania Singers group at Max Planck Institute in Leipzig (Klimecki et al. 2014).

About 60 volunteers were divided into two groups. One group meditated on love and compassion, and the others were instructed to cultivate feelings of empathy for others. After a week of compassion meditation, participants had more positive feelings while looking at video clips with suffering people. The other group who had solely cultivated empathy, experienced emotions that resonated with the sufferings of the subjects shown in the video clips, but these shared emotions also resulted in negative feelings and thoughts, and the empathy cultivation group exhibited significant more distress.

The compassion practice entailed noticing suffering in others (and one-self) and a cultivating of attitudes and feelings of loving kindness and compassion toward other people, whether they are close relatives, strangers or enemies, as well as to receive compassion from others, and for one self.

---

[58]This definition also differentiates compassion from empathy, which refers to the vicarious experience of another's emotions. Lazarus R. (1991). *Emotion and adaptation* Oxford University Press. However, we acknowledge that research on compassion is complex, not least due to the different scientific understandings of the topic. For an excellent overview, see Goetz, J. L., Keltner, D., & Simon-Thomas, E. (2010). Compassion: an evolutionary analysis and empirical review. *Psychological bulletin*, *136*(3), 351.

[59]Hofmann, S. G., Sawyer, A. T., Witt, A. A., & Oh, D. (2010). The effect of mindfulness-based therapy on anxiety and depression: A meta-analytic review. *Journal of consulting and clinical psychology*, *78*(2), 169.

[60]Gilbert, P. (2010). Compassion focused therapy: Distinctive features. New York, NY: Routledge.

[61]Bellini, L. M., Baime, M., & Shea, J. A. (2002). Variation of mood and empathy during internship. *Jama*, *287*(23), 3143–3146.

Research suggests that a specialized affect regulation system (or systems) underpins feelings of reassurance, safeness and well-being. It is believed to have evolved with attachment systems and, in particular, the ability to register and respond with calming and a sense of well-being to being cared for.[62] This affect regulation system is poorly accessible in people with high shame and self-criticism, in whom the 'threat' affect regulation system dominates orientation to their inner and outer worlds. When psychology professor Paul Gilbert noticed that his patients had a lot of shame, self-criticism and self-attacking he responded by developing a therapy that draws on evolutionary, social, developmental and Buddhist psychology, and neuroscience, which is called Compassion focused therapy.[63] The aim is to help people develop and work with experiences of inner warmth, safeness and soothing, via compassion and self-compassion. It should be noted that this type of self-compassion is distinctly different from e.g. self-esteem, or pity.[64, 65] The skill we want to highlight here is a robust, trainable and stable capacity, which in its most pure form is not particularly vulnerable to external circumstance (e.g. other people's opinion, trends in society, social pressure etc.). It has to do with identity and self-awareness. It is in-ward looking—I am both capable and willing to accept what type of person I am (my potential and my limitations) but it is also outward looking as it also relates and regulates how I am interconnected with other beings. This idea fits well with some interpretations of virtue ethics on which virtue ethics contains a distinctly forward looking element. Very briefly the idea is as follows; because I know who I am I can also contemplate my personal development (in relation to my environment) i.e. what type of person I aspire to be and how to get there.[66]

---

[62]Gilbert, P. (2009). Introducing compassion-focused therapy. *Advances in Psychiatric Treatment* 15: 199–208.

[63]"Compassion focused therapy (CFT) is rooted in a functional analysis of basic social motivational systems. During human evolution a range of cognitive competencies for reasoning, reflection, anticipating, imagining, mentalizing evolved, as well as a socially contextualized sense of self. These new competencies can cause major difficulties in the organization of (older) motivation and emotional systems. Our evolved brain is therefore potentially problematic because of its basic 'design,' being easily triggered into destructive behaviours and mental health problems. CFT highlights the importance of developing people's capacity to (mindfully) access, tolerate, and direct affiliative motives and emotions, for themselves and others, and cultivate inner compassion as a way for organizing ourselves in prosocial and mentally healthy ways". Gilbert, P. (2010). *Compassion focused therapy: Distinctive features.* New York, NY: Routledge. Gilbert, P., Clark, M., Hempel, S., Miles, J.N.V. & Irons, C. (2004) Criticising and reassuring oneself: An exploration of forms, styles and reasons in female students. *British Journal of Clinical Psychology,* 43, 31–50.

[64]Neff, K. D., & Vonk, R. (2009). Self-compassion versus global self-esteem: Two different ways of relating to oneself. *Journal of personality*, *77*(1), 23–50.

[65]Smeets, E., Neff, K., Alberts, H., & Peters, M. (2014). Meeting Suffering With Kindness: Effects of a Brief Self-Compassion Intervention for Female College Students. *Journal of clinical psychology*, *70*(9), 794–807.

[66]This will be expanded on in Chap. 5 but the reader should note already here that this line of argument does not imply that the virtues are instrumental to the good life. Rather, there is a little means and a lot of ends in all the virtues.

2.7 Some Challenges with Meditation Practice and Research

So far we have presented a rather bright picture of the possibilities with meditation practices. Evidently, there are also many problems attaching both to practice and how to measure results in a scientific way. Below follows a discussion on some key challenges with meditation practice and meditation research.

2.7.1 *Small Groups*

In applied meditation research, participants are often randomized to either a meditation training group or a "wait-list" control group in which there is no training. Both groups undergo comparable testing before and after the period of time taken for the meditation training condition. It is of importance to remember that the studies we cite are based on group statistics: even when there is a strong positive outcome in favor of the meditation group, not every single individual in the meditation group experiences a positive response. The differences between groups are often not that large.

2.7.2 *Not so Blind*

Furthermore, participants can never be truly blind to the condition in which they have been randomized. Those in the training condition are likely to be biased toward different *expectations* e.g. the potential benefit (or lack thereof).[67] In addition, they might (often subconsciously) be motivated to seek to please the teacher. It is not possible to disentangle such bias from the specific effects of the meditation practice itself and, unfortunately, many of the available experimental studies of meditation in applied settings have employed this kind of research design. That said, alternative designs are of course possible. One good example is studies where participants are randomly assigned to a meditation training condition versus another training condition, such as physical exercise, didactic instruction, or simple muscle relaxation training. Meditation and other active training condition can be carefully equated for such things as amount of training, amount of home practice, credibility and enthusiasm of instructors. These kinds of randomized

[67]This was described by Landsberger already in 1950. The Hawthorne effect is a phenomenon in which individuals improve an aspect of their behavior in response to their awareness of being observed. Henry A. Landsberger. (1958). *Hawthorne Revisited*, Ithaca. McCarney R, Warner J, Iliffe S, van Haselen R, Griffin M, Fisher P; Warner; Iliffe; Van Haselen; Griffin; Fisher (2007). "The Hawthorne Effect: a randomised, controlled trial". *BMC Med Res Methodol* **7**: 30.

active control research designs are being increasingly employed and yielding critically important new data.[68]

2.7.3 *Decreased Well-Being*

It is also important to remember that it is highly unlikely that someone who signed up for a meditation class and joined all of these various studies cited above would actually demonstrate all of the benefits seen across all of the different meditation groups being studied. Just in terms of the law of averages, we would actually expect that some people in the meditation group not to benefit from meditation and others might even experience a temporarily increase in anxiety, or have their performance decreased by the practice.

Negative side-effects of meditation practice are very seldom reported, but there is an increasing awareness of such risks.[69] There might be some publication bias towards positive effects, as there are several small published studies, which might be biased towards the intervention being tested. The question of opportunity cost (i.e. the time spent in meditation might outcompete other valuable, virtuous activities) is rarely mentioned in the published studies.

2.7.4 *Other Issues*

These are far from the only issues that contemplative science has to deal with. Additional challenges include: How best to assess the subjective experience of meditation practitioners? How to define and measure mindfulness? How to define and measure compassion? How to overcome the inherent biases in self-report measures, and how to best interpret data from modern brain imaging/neurophysiology methods?[70]

On a brighter note, there are, as seen in this chapter well-documented effects of both mindfulness and compassion practices for both healthy subjects struggling

---

[68]Kazniak A. 2014 Huffington post. www.huffingtonpost.com/al-kaszniak/progress-and-challenge-in_b_6083478.html.

[69]Dimidjian, S., & Hollon, S. D. (2010). How would we know if psychotherapy were harmful? *American Psychologist*, *65*(1), 21. This study examines examples where psychotherapy has caused serious harm to a patient, and highlights the value of creating standards for defining and identifying when and how harm can occur at different points in psychotherapy. It has inspired research on side effects of meditation, which is now increasingly investigated and e.g. reported in this article: Compson, J. (2014). Meditation, Trauma and Suffering in Silence: Raising Questions about How Meditation is Taught and Practiced in Western Contexts in the Light of a Contemporary Trauma Resiliency Model. *Contemporary Buddhism*, (ahead-of-print), 1–24.

[70]Kerr, C. (2014). Don't believe the hype. Interview in Tricycle, October 01, 2014 www.tricycle.com/blog/don't-believe-hype.

with decisions-making and other "normal" life challenges, as well as in people with more severe symptoms of common public health issues such as anxiety, depression and chronic pain. The fact that these behavioral changes were accompanied by observable changes in brain function and structure (as summarized above) further supports the hypothesis that is defended here.

To sum up this chapter: The function and structure of the adult brain changes in response to e.g. external environmental impacts, and this phenomenon is called neuroplasticity. Central to our cognitive functioning is attention—the challenge to sustain it comes often from both external an internal distractions. Mind-wandering is frequently described as a bad thing, as it disenables focused attention, but could be beneficial if one is priming the mind-wandering episode with e.g. interesting musings. Juggling, playing and rehearsing music, driving taxi in a complex city, meditation (e.g. Attention Training Meditation, Open Monitoring and Compassion Training) can be trained and will also produce changes to both the brain and behavior.

In Chap. 3 we will discuss some alternative methods for improving cognitive capacities, such as pharmaceuticals and certain types of technology that can be used to improve our cognitive skills, and what risks that might attach to such practices.

In Chap. 4 we take a closer look at cognitive flexibility. Firstly, we define this core cognitive capacity and explain why it is good to have it to a high degree. Secondly, we examine the link between the meditation techniques promoted in this chapter and increased cognitive flexibility. Thirdly, we point out that high cognitive flexibility does not guarantee responsible moral decision-making. Consequently we need a robust, and action guiding, moral framework which can anchor these capacities and guide vacillating agents. Chapter 4 finishes with a brief discussion of the connection between improved core cognitive capacities and the installing of a set of key epistemic virtues.

References

Britton, W. B., Lindahl, J. R., Cahn, B. R., Davis, J. H., & Goldman, R. E. (2014). Awakening is not a metaphor: The effects of Buddhist meditation practices on basic wakefulness. *Annals of the New York Academy of Sciences, 1307*(1), 64–81.

Goleman, D. (2013). *Focus: The hidden driver of excellence*. A&C Black.

James, W. (1890). *The principles of psychology, attention*, Chapter 11, 404.

Klimecki, O. M., Leiberg, S., Ricard, M., & Singer, T. (2014). Differential pattern of functional brain plasticity after compassion and empathy training. *Social Cognitive and Affective Neuroscience, 9*(6), 873–879.

Mischel, W. (2014). *The marshmallow test: Understanding self-control and how to master it*. Random House.

Zajonc, A. (2009). *Meditation as contemplative inquiry: When knowing becomes love*. Great Barrington, MA: Lindisfarne Books.

Chapter 3
The Methods

Abstract Chapter 1 introduced the key ideas and identified the need for cognitive improvements. The first half of Chap. 2 consisted of a schematic review of the neurophysiological background for the development of such capacities. It was, for example, explained that the adult brain is plastic enough for both functional and structural changes to take place. The latter half of Chap. 2 was devoted to exploring three types of mental training techniques which, in evidence based studies, have been shown to have a positive, lasting and generalizable effect on cognitive capacities. Evidently these techniques are not without problems, nor are they the only methods around. The present chapter discusses some alternative methods for improving cognitive capacities. It includes an introductory account of how some pharmaceuticals and certain types of technology can be used to improve our cognitive skills and what risks that might attach to such practices.

Keywords Mental training · Neuroenhancement · Pharmaceutical drugs · Technology · Life-style

3.1 How to Improve

The overarching purpose of this book is to explore how improved cognitive and emotional regulation skills (our examples include a set of core capacities such as cognitive flexibility, increased focus/attention and meta-awareness, "controlled" mind-wandering as well as improved capacity for emotional regulation), might enable better (as in more responsible and less biased) decision-making.

In the last chapter we described three specific training techniques i.e. Focused Attention Training, Open awareness and mindfulness training and Compassion Training. We assume that all three techniques have concrete, most of them lasting, structural and functional changes in the human brain, of a kind that plausibly could be taken to positively impact our cognitive capacities. Documented examples include:

B. Fröding and W. Osika, *Neuroenhancement: How Mental Training and Meditation Can Promote Epistemic Virtue*, SpringerBriefs in Ethics,
DOI 10.1007/978-3-319-23517-2_3

improved focus,[1] alertness[2] and working memory[3] emotional stability (equanimity),[4] improved capacity for introspection,[5] reduced stress and misplaced fear, adequate compassionate behavior[6] and improved sense of impartiality. We also speculated that those skills could be generalizable outside the original training scenario.

As shown, there is good scientific evidence that regular meditation enables you to focus your attention and reduces the level of general stress in the body. Unfortunately, however, there is no automatic connection between reduced attention diversion and better decision-making in the moral sense of the word. As previously explained, 'better' in this context means more than merely efficient and on our account good decision-making includes a moral component and a propensity towards pro-social behaviour. So, while studies have shown that regular meditation (in most cases) improves the individual's attention and focus (both inward and outward) and reduce stress, it cannot be taken for granted that the person will also become morally better. The skill you acquire is morally neutral and needs to be complemented by a moral framework which is action guiding and helps the agent to put the cognitive capacities to good use. We will return to discuss such aspect at the end of Chap. 4.

3.2 Defining Cognitive Enhancement

If we chose to follow the broad definition of cognitive enhancement offered by Juengst[7] (i.e. as inventions in humans that aim to improve mental functioning beyond what is necessary to restore good health) then it is clear that non-pharmacological ways to improve our mental capabilities also count as cognitive enhancement. This is

---

[1]Lutz, A., Slagter, H. A., Dunne, J. D., & Davidson, R. J. (2008). Attention regulation and monitoring in meditation. *Trends in cognitive sciences*, *12*(4), 163–169.

[2]Britton, W. B., Lindahl, J. R., Cahn, B. R., Davis, J. H., & Goldman, R. E. (2014). Awakening is not a metaphor: the effects of Buddhist meditation practices on basic wakefulness. *Annals of the New York Academy of Sciences*, *1307*(1), 64–81.

[3]Jha, A. P., Stanley, E. A., Kiyonaga, A., Wong, L., & Gelfand, L. (2010). Examining the protective effects of mindfulness training on working memory capacity and affective experience. *Emotion*, *10*(1), 54.

[4]Ricard, M., Lutz, A., & Davidson, R. J. (2014). Mind of the Meditator. *Scientific American*, *311* (5), 38–45.

[5]Kerr, C. E., Josyula, K., & Littenberg, R. (2011). Developing an observing attitude: an analysis of meditation diaries in an MBSR clinical trial. *Clinical Psychology & Psychotherapy*, *18*(1), 80–93.

[6]Smeets, E., Neff, K., Alberts, H., & Peters, M. (2014). Meeting Suffering With Kindness: Effects of a Brief Self-Compassion Intervention for Female College Students. *Journal of clinical psychology*, *70*(9), 794–807. Gilbert, P. (2010). Compassion focused therapy: Distinctive features. New York, NY: Routledge.

[7]Juengst, E. T. (1998). What does enhancement mean. *Enhancing human traits: Ethical and social implications*, 29–47.

an important point as much of the current debate is limited to enhancement through drugs and/or technology. And, related to this, that there is a tendency towards using sci-fi examples which explore the ethical aspects of counterfactual scenarios rather than the issues we are faced with today or in the near future.

This is unfortunate both because it tends to alienate people and increase the distance between science and society and because that leaves the current issues unaddressed or at least neglected.

As pointed out in a recent article by Dresler et al.,[8] there are indeed a great number of potentially powerful non-pharmacological tools and techniques available. The methods are split into two broad groups. The first one includes things that all humans do simply to survive and thrive; nutrition, physical exercise and sleep and the second includes e.g. mnemonic strategies, computer training and brain stimulation. The authors explore and evaluate the effects of these and other non-pharmacological ways in which we can achieve cognitive improvements and enhancements.

As should be clear to the reader by now we are sympathetic to the broader definition above. Consequently, we are interested in examining non-pharmacological strategies and more precisely life-style choices in the form of mental training and meditation techniques to see how that can improve some core cognitive capacities in a long-lasting and generalizable manner. While showing great promise, researching such techniques and documenting their effects in scientifically sound way is not without challenge. Chapter 2 ended with a discussion of some of the main problems attaching to meditation techniques as methods for cognitive enhancement. Problematic aspects include the probability of a certain publication bias towards positive effects.

In addition, participants often are randomized to either a meditation training condition or a "wait-list" control condition. The latter group gets no training. While both groups are subjected to comparable testing before and after the period of time taken for the meditation training condition, the very design of the study clearly makes it hard to tell if it was the meditation practice per se, that caused the observed changes. Related to this, those in the training condition are likely to be biased toward different *expectations* for benefit and such motivations as wanting to please the researcher, compared to those in the wait-list control condition.[9]

Another challenge is that the studies are based on group statistics. Even when there is a strong positive outcome in favor of the meditation group, not every single individual in the meditation group would have experienced a positive response.

---

[8]Dresler, M., Sandberg, A., Ohla, K., Bublitz, C., Trenado, C., Mroczko-Wąsowicz, A., … & Repantis, D. (2013). Non-pharmacological cognitive enhancement. *Neuropharmacology*, *64*, 529–543.

[9]Landsberger, H. A. (1958). Hawthorne Revisited: Management and the Worker, Its Critics, and Developments in Human Relations in Industry. See also CH *2.3.4.2. Not so blind.*

There is also an increasing awareness of side effects of meditation in the contemplative science community.[10] The question of opportunity cost (i.e. the time spent in meditation might outcompete other valuable, virtuous activities) is rarely mentioned in the published studies. Other examples of remaining questions include: How best to assess the subjective experience of meditation practitioners, overcome the inherent biases in self-report, and integrate them with other measures such as brain imaging? How to define and measure mindfulness, and improved introspective skills? How to define and measure compassion?[11]

We are of the opinion that many of these problems can be overcome and that the mental training and meditation techniques explored here have some distinct advantages. To better assess the potential of regular meditation it might, however, be helpful to take a brief look at some of the alternatives. Below follows a set of concrete examples investigating the enhancing effects of oxytocin, serotonin and brain machine interface technology (BMI). Towards the end of the chapter we return to our favored techniques and discuss how they can be combined with other forms of training methods e.g. especially developed computer games. While using computer games as an aid for improving cognitive skills is fairly new it has massive potential and has, not surprisingly perhaps, attracted a lot of interest recently. Given the rate at which technology develops it seems very likely that such games will become both more efficient and more readily available to the many within the next decade.[12] We will argue that there is no necessary tension between the two methods and, further, that combining them (both as in engaging in both and as in developing highly advanced games and apps which reinforce the effects of the mental training and meditation to create a positive spin which reinforces and perpetuates the positive behavior).

[10]Dimidjian, S., & Hollon, S. D. (2010). How would we know if psychotherapy were harmful? *American Psychologist*, *65*(1), 21. Compson, J. (2014). Meditation, Trauma and Suffering in Silence: Raising Questions about How Meditation is Taught and Practiced in Western Contexts in the Light of a Contemporary Trauma Resiliency Model. *Contemporary Buddhism*, (ahead-of-print), 1–24.

[11]"It's important to understand that (1) brain imaging cannot do away with the basic gap between subjective experience and objective measurement (2) complex subjective experiences like those felt in meditation are likely made up of a complex array of brain mechanisms that cannot be captured by the simple sets of hypotheses that can be tested in a single brain experiment." Interview with Catharine Kerr, by Knapp, A, retrieved from Forbes on line 9/09/2011.

[12]To get an idea of the pace consider e.g. Moore's law which (correctly so far) stipulates that the number of integrated circuits double every 24 months. Moore, G. E. (1965). Cramming more components onto integrated circuits, 1965. *Electronics Magazine*, 4.

3.3 Three Methods

Broadly speaking, there are three main ways in which we can influence our cognitive capacities (i) pharmacological cognitive enhancers,[13] hormones, neurotransmitters and nootropics,[14] (ii) technology and (iii) life-style (including e.g. diet and physical as well as mental training). Evidently all three methods can have a positive as well as a negative impact and there are risks involved in all three.

In this section we will compare and contrast the following methods for cognitive enhancement; oxytocin (a hormone), selective serotonin reuptake inhibitors (as pharmaceutical drugs described to modify serotonin levels), wearable technology and Brain Machine Interface technology (BMI). Before setting off, it should be noted that we by no means wish to imply that these and other methods are without merit—we are certainly not out to discredit them. Quite to the contrary both hormones and pharmaceutical drugs might, if the risks and negative side effects are compensated for to an adequate extent, become very useful. As for technology, given the fast pace of development and level of integration in almost all aspects of our lives it would be surprising if the research into e.g. BMI and neuro-prosthetics as broadly conceived did not yield some important insights in the relatively near future. Regarding the uptake levels of said technology one can of course only speculate. The fact that most people, through sheer necessity and habituation, have developed a relatively tolerant—if not out-rightly positive—attitude to technology might be taken as an indication that uptake could be high. Further to this, the fields of different brain stimulation techniques[15] as well as of neuro-feedback methods are receiving much interest.

3.4 Pharmaceuticals, Hormones and Neurotransmitters

As previously pointed out (see e.g. Chap. 1) our focus on life-style does not imply that we believe that popular pharmaceuticals like methylphenidate (e.g. Ritalin®), amphetamine (Adderall®), dopamine agonists (e.g., Mirapex®), acetylcholine esterase inhibitors (e.g., Donepezil®), modafinil (Provigil®) or hormones cannot be effective for achieving cognitive improvements. Nor do we wish to argue that

---

[13]Pharmacological cognitive enhancers are used to treat neurodegenerative (e.g. Alzheimer's and dementia) and neuropsychiatric disorders (e.g. schizophrenia and ADHD).

[14]The word *nootropic* is derived from the Greek words *nous*, or "mind", and *trepein* meaning to bend or turn. Nootropics are drugs, supplements, nutraceuticals, and functional foods that improve one or more aspects of mental function, such as working memory, motivation, and attention. Giurgea, C. (1972). Pharmacology of integrative activity of the brain. Attempt at nootropic concept in psychopharmacology. *Actualités pharmacologiques*, *25*, 115.

[15]Bennabi, D., Pedron, S., Haffen, E., Monnin, J., Peterschmitt, Y., & Van Waes, V. (2014). Transcranial direct current stimulation for memory enhancement: from clinical research to animal models. *Frontiers in systems neuroscience*, *8*. Clark, V. P., & Parasuraman, R. (2014). Neuroenhancement: enhancing brain and mind in health and in disease. *Neuroimage*, *85*, 889–894.

research and drug trials should be stopped. This deserves pointing out since such research is criticized and referred to as "cosmetic neurology".[16] At present, however, their attractiveness[17] is undermined as the effects of drugs like Provigil and Ritalin on the normal brain—i.e. when people use them for non-therapeutic effects seeking to increase concentration or to combat fatigue—are small, if even measurable. In addition, it seems that the drugs only affect single cognitive capacities/functions as opposed to e.g. overall performance or even intelligence. Previous research raises the possibility that modafinil substances do not improve the performance of certain neuropsychological tasks, because these tasks involve not only wakefulness and attentional components, but also sophisticated problem-solving abilities which modafinil may not be able to enhance.[18]

Further, and much more seriously, the long-term risks and negative side-effects for the individual (e.g. potential for addiction in the case of methylphenidate and modafinil, especially when used outside the initially intended conditions such as severe problems with attention deficits and hyperactivity) needs more research.[19] As for the potential of negative effects in the form of harm to others e.g. negative socio-economic effects, lack of fairness, undermining of equity and uneven distribution please see Chap. 6.

In the last 10–15 years the hormone oxytocin, also called 'the love hormone' and pharmaceuticals that affect levels of the neurotransmitter serotonin such as selective serotonin reuptake inhibitors[20] have received much attention. Some have hailed

---

[16]Chatterjee, A. (2004). Cosmetic neurology The controversy over enhancing movement, mentation, and mood. *Neurology, 63*(6), 968–974.

[17]Maher, B. (2008). Poll results: look who's doping. *Nature, 452*, 674–675. Repantis, D., Schlattmann, P., Laisney, O., & Heuser, I. (2010). Modafinil and methylphenidate for neuroenhancement in healthy individuals: a systematic review. *Pharmacological Research, 62*(3), 187–206.

[18]Mohamed, A. D., & Lewis, C. R. (2014). Modafinil Increases the Latency of Response in the Hayling Sentence Completion Test in Healthy Volunteers: A Randomised Controlled Trial. *PLoS ONE, 9*(11). Mohamed AD (2014) The Effects of Modafinil on Convergent and Divergent Thinking of Creativity: A Randomized Controlled Trial. *The Journal of Creative Behavior.*

[19]Despite being on the market for several decades (methylphenidate was synthesized in 1944 and identified as a stimulant in 1954), the tolerance and mechanisms of action of e.g. methylphenidate are still under study. In view of the questions that remain unanswered regarding tolerance of this drug in the long term, the common recommendation is that its use should be strictly reserved to the approved indications, with evaluation of the risk-benefit ratio on a case-by-case basis. T., Herlem, E., Taam, M. A., & Drame, M. (2014). Methylphenidate off-label use and safety. *SpringerPlus, 3* (1), 286. There is however a debate regarding the acceptability of using such substances from a fairness perspective see e.g. Harris J (2009). "Is it acceptable for people to take methylphenidate to enhance performance? Yes". *BMJ* 338: b1955. Chatterjee A (2009). "Is it acceptable for people to take methylphenidate to enhance performance? No". BMJ 338: b1956.

[20]Kirsch, I., Moore, T. J., Scoboria, A., & Nicholls, S. S. (2002). The emperor's new drugs: an analysis of antidepressant medication data submitted to the US Food and Drug Administration. *Prevention & Treatment, 5*(1), 23a. Fountoulakis, K. N., & Möller, H. J. (2011). Efficacy of antidepressants: a re-analysis and re-interpretation of the Kirsch data. *International Journal of Neuropsychopharmacology, 14*(3), 405–412.

them as key to making humans more trusting and empathic and less prone to 'in-group thinking' and aggression. In other words: that modifying the levels of them would bring about moral improvement. While both oxytocin and serotonin indeed have a strong impact on behavior both in animals and humans it is less clear that it is always for the better. Recent research have raised some warning flags, or at least painted a more nuanced picture.[21] It would appear that, as so often in life, more of a good thing is not very good at all.

3.4.1 Oxytocin

On discovering that the hormone oxytocin reduces aggressive behavior and increases pro-social behavior it became known as 'the love hormone'.[22] The interest started with a number of studies conducted on prairie voles and their sexual behavior. Very briefly it turned out that by modifying the levels of oxytocin (in the female prairie vole) and vasopressin (in the male prairie vole) the researchers could affect the animals' tendency for pair bonding and how promiscuously the voles behaved.[23] The next step was to explore what, if any, the implications for human behavior and cognitive enhancement might be. Several studies reported positive results indicating that the level of oxytocin also could contribute to explaining human love, partner bonding and parental love.[24]

These and other research results in the same vein gave rise to speculations of 'love drugs' which could have a positive impact on love and attachment and make people more prone to both form, and remain in, relationships.[25] The research also showed, however, that the less promiscuous the male voles became the more

---

[21]For example reinforcing in-group bias and aggressive behavior in the case of oxytocin and becoming overly trusting in the case of serotonin.

[22]For a review see Heinrichs, M. et al. (2009) Oxytocin, vasopressin, and human social behavior. *Front. Neuroendocrinol.* 30, 548–557. Macdonald, K. and Macdonald, T.M. (2010) The peptide that binds: a systematic review of oxytocin and its prosocial effects in humans. *Harv. Rev. Psychiatry* 18, 1–21.

[23]Cho, M.M., A.C. DeVries, J.R. Williams, and C.S. Carter. (1999). The effects of oxytocin and vasopressin on partner preferences in male and female prairie voles (Microtus ochrogaster). *Behavioral Neuroscience* 113(5): 1071–1079. Harbaugh, and T.R. Insel. (1993). A role for central vasopressin in pair bonding in monogamous prairie voles. *Nature* 365(6446): 545–548.

[24]Wang Z, Young L.J., De Vries G.J., Insel T.R. 1998. Voles and vasopressin: a review of molecular, cellular, and behavioral studies of pair bonding and paternal behaviors. Prog Brain Res. 119: 483–99. Nair, H.P., and L.J. Young. 2006. Vasopressin and pairbond formation: genes to brain to behavior. Physiology 21: 146–152.

[25]Kosfeld, M., Heinrichs, M., Zak, P. J., Fischbacher, U., & Fehr, E. (2005). Oxytocin increases trust in humans. *Nature, 435*(7042), 673–676. Savulescu, J., & Sandberg, A. (2008). Neuroenhancement of love and marriage: the chemicals between us. *Neuroethics, 1*(1), 31–44. Schneiderman, I., Zagoory-Sharon, O., Leckman, J. F., & Feldman, R. (2012). Oxytocin during the initial stages of romantic attachment: Relations to couples' interactive reciprocity. *Psychoneuroendocrinology,* 37, 1277–1285.

aggressively and possessively they behaved around their partners. Perhaps not surprisingly, it would appear that the connections and effects are more complex than what was initially hoped for. It seems that oxytocin impacts a range of behavior, some of which is positive and pro-social and other, which clearly are not. In humans, some negative effects are e.g. increased inclination toward interpersonal *violence*, especially in those who have higher trait aggression.[26] It can impair or even hinder trust building and cooperation in vulnerable individuals,[27] and can also lead to an increased tendency to feel envy and schadenfreude.[28] Hence, it is becoming more evident that context and even the genetic make-up matter for how the effects of oxytocin are expressed, and that these effects not are restricted to 'increased levels of love' as broadly understood.[29]

3.4.2 Oxytocin as a Method for Moral Enhancement

In her seminal work on neuroscience and moral decision-making Professor Pat Churchland points out that the potential for using oxytocin on humans in order to alter and improve behavior and actions (i.e. as a moral enhancer)[30] seems to be subject to much and even to pure speculation—hormones work in complex ways. Churchland's broader position seems to be in line with the claims made here. For example; our morality and capacity for good decision-making is shaped by an array of factors (some of which are part of our biological make-up, others which are related to our emotions and others which are to do with environmental factors etc.). In other words—that while the neural platform might well be the base for our morality it does not tell the whole story. Consequently then, there are plenty of

---

[26]DeWall, C. N., Gillath, O., Pressman, S. D., Black, L. L., Bartz, J. A., Moskovitz, J., & Stetler, D. A. (2014). When the Love Hormone Leads to Violence Oxytocin Increases Intimate Partner Violence Inclinations Among High Trait Aggressive People. *Social Psychological and Personality Science*, 1948550613516876.

[27]Bartz, J., Simeon, D., Hamilton, H., Kim, S., Crystal, S., Braun, A., … & Hollander, E. (2010). Oxytocin can hinder trust and cooperation in borderline personality disorder. *Social Cognitive and Affective Neuroscience*, nsq085.

[28]Shamay-Tsoory, S. G., Fischer, M., Dvash, J., Harari, H., Perach-Bloom, N., & Levkovitz, Y. (2009). Intranasal administration of oxytocin increases envy and schadenfreude (gloating). *Biological psychiatry*, *66*(9), 864–870.

[29]Bartz, J. A., Zaki, J., Bolger, N., & Ochsner, K. N. (2011). Social effects of oxytocin in humans: context and person matter. *Trends in cognitive sciences*, *15*(7), 301–309. Declerck, C.H. et al. (2010) Oxytocin and cooperation under conditions of uncertainty: the modulating role of incentives and social information. *Horm. Behav*. 57, 368–374.

[30]Professor Churchland, University of California San Diego, is a leading researcher in this field and has published extensively on neuroscience, consciousness, free will etc. See e.g. Churchland, P. S. (2011). *Braintrust: What neuroscience tells us about morality*. Princeton University Press. http://philosophyfaculty.ucsd.edu/faculty/pschurchland/index_hires.html.

behavioral ways (which are not only available and risk tested but, often, quite efficient) which can help most people change and improve.[31]

Indeed there is no shortage of examples that indicate that we can promote a desired behavior by working hard and practice. In addition to improving our propensity to behave morally such techniques might also improve our moral judgment in a way that is generalizable to situations outside the immediate learning scenario, which in neuroscience is called "transfer effects". Evidently to have good moral judgment involves having a dynamic capacity as opposed to a skill which is simply learnt by heart. That said, many such skills are of course highly useful in other domains e.g. car-driving or sailing or ice-skating. Morality, on the other hand, is perhaps more analogous to mastering a foreign language. Learning the grammar and developing a substantive vocabulary might well make you fluent but you will not be on a par with a native speaker (see Chaps. 4 and 5 for more on the necessity of a moral framework which can ensure that the cognitive improvements result in responsible decision-making).

Given what we know today, i.e. that while oxytocin can make us more prone to cooperate it can also make us more jealous and aggressive, it is too blunt an instrument for bringing about what one might call general good moral judgment. While it might have a positive effect in some situations, some of the time it is difficult to see how it can be conducive to shaping a dependable capacity for moral judgments. In comparison then, life-style changes appear a more prudent strategy for altering behavior than tweaking the levels of oxytocin.[32] While it might be less efficient and require much harder work we simply do not know enough about the overall and long-term effects of using hormones and neurotransmitters in this way and are consequently not well placed to assess the risks and the wisdom of using oxytocin.

This lack of knowledge might make us disregard the possibility of the unknown unknowns or Black Swan events as they are sometimes called. Further, we might also be lead to attribute to little weight to the known unknowns and thus fail to identify and allocate resources to various risk scenarios (for more on this see Chap. 1).

---

[31]Philosophy Bites interview, August 3rd, 2012, Pat Churchland on What Neuroscience Can Teach Us About Morality, http://philosophybites.com/2012/08/pat-churchland-on-what-neuroscience-can-teach-us-about-morality.html.

[32]It should be noted that we by no means wish to imply that environmental factors are 'weak' when it comes to shaping behavior. Quite to the contrary there are plenty of examples showing that they can be life altering both for the good and the bad. Christakis, N. A., & Fowler, J. H. (2007). The spread of obesity in a large social network over 32 years. *New England journal of medicine*, *357*(4), 370–379. Fowler, J. H., & Christakis, N. A. (2008). Dynamic spread of happiness in a large social network: longitudinal analysis over 20 years in the Framingham Heart Study. *Bmj*, *337*, a2338. Rappaport, S. M. (2012). Discovering environmental causes of disease. *Journal of epidemiology and community health*, *66*(2), 99–102.

3.4.3 Serotonin and Selective Serotonin Reuptake Inhibitors (SSRI)

Very briefly it could be said that serotonin is a neurotransmitter primarily found in the gastrointestinal tract, platelets and the central nervous system. It is often described as a contributor to the regulation of feelings of well-being and happiness.[33]

In the central nervous system it is synthesized in serotonergic neurons, where it has various functions including the regulation of mood, appetite, and sleep. Serotonin is also involved in cognitive functions such as memory and learning. Modulation of serotonin at synapses is thought to be a major action of several classes of pharmacological antidepressants, such as selective serotonin re-uptake inhibitors (SSRI). The effect of SSRI is thought to depend on the production of positive biases in the processing of emotional information. Both behavioral and neuroimaging studies show that SSRI administration produces positive biases in attention, appraisal and memory from the earliest stages of treatment, well before the time that clinical improvement in mood becomes apparent.[34]

Modification of the serotonin equilibrium has also been shown to affect moral behaviour. In a series of experiments Crockett[35] highlighted that the use of citalopram (a SSRI) impacts our level of aversion to unjust treatment and that high levels tend to make people (more) prone to cooperate. The effect was differentiated in the way that individuals high in trait empathy showed stronger effects of citalopram on moral judgment and behaviour than individuals low in trait empathy. In her research Crockett has, amongst other things, sought to identify the mechanisms that regulate our propensity for collaboration and how to balance, and reduce, some of the biological reactions, which might lessen such pro-social behaviour.

So far the observable effects of SSRIs on our moral behavior are very small. As indicated above, people who scored high on empathy before taking the medication tend to display a stronger aversion to unjust treatment/resource allocation in the experimental game situations (also if it is advantageous to themselves). They seem to (at least some of the time) prefer fairness to getting the bigger share of the money in the experiments. Perhaps one can say that it tends to reinforce positive behavior in the already good ones rather than improve the bad people. Then again, there are also some reports that people who are under the influence of serotonin modifying medication tend to be less prone to punish people who cheat in the ultimatum game experiments.

---

[33]Young SN (2007). "How to increase serotonin in the human brain without drugs". *Rev. Psychiatr. Neurosci.* 32 (6): 394–399.

[34]Harmer, C. J., & Cowen, P. J. (2013). 'It's the way that you look at it'—a cognitive neuro-psychological account of SSRI action in depression. *Philosophical Transactions of the Royal Society B: Biological Sciences, 368*(1615), 20120407.

[35]Professor Crockett a neuro-scientist at UCL has published extensively on morality, altruism and decision making.

If correct such an effect could prove quite negative for the player as she would accept wrongful behavior, free riders and thus compromise fairness and just allocation. We have already mentioned a similar phenomenon when investigating the effects of tranquilizers, e.g. benzodiazepines on fairness.[36] Evidently, treatment with SSRI and other pharmaceuticals that modulate serotonin levels also have side effects, some of which are really serious (but fortunately also very rare).[37]

Awaiting a breakthrough on the science front with regards to the more precise effects and potential benefits of various neurotransmitters, we can still learn a lot from the research of Professor Crockett and her colleagues. One interpretation of these experiments is that we ought to look closely at the role of embedding structures and their impact on people's behavior.

It is clear that certain situations and environments trigger responses and actions (e.g. misplaced aggression and fear) which are not very conducive to our overall well-being (not as individuals nor as a society). The experiments indicate that some of these responses can be explained by our biological make-up and that they potentially can be mitigated or reduced by the levels of neurotransmitters and hormones in the brain. It appears that our biology and our environment conspire and lead us to make bad choices, but at the moment we do not have a very clear picture of how this can be improved from a neurochemical point of view.

However, as we learn more about the biological explanations for our cognitive limitations (and the reactions that follow) we can seek to balance some of the negative outcomes through a combination of life-style and social organization. This opens for the rather hopeful prospect that we can change and, consequently, that we are not slaves to our biology in the base instincts kind of way. There is a host of good examples of how people can be socialized and civilized[38] and behave better in a way that encourages and promotes such behavior (for more on this see Chaps. 5 and 6).

3.5 Technology

3.5.1 Wearables and Brain Machine Interface BMI

Another route to enhancement comes through wearing technological devices. Prime examples include sensors and actuators enabling increased body awareness and smartphone applications which facilitate (and remind the user of) mental training

---

[36]See Chap. 1, Sect. 1.4.2 page 10. Gospic K, Mohlin E, Fransson P, Petrovic P, Johannesson M, et al. (2011) Limbic Justice—Amygdala Involvement in Immediate Rejection in the Ultimatum Game. *PLoS Biol* 9(5): e1001054.

[37]Extremely high levels of serotonin can cause a condition known as serotonin syndrome, with toxic and potentially fatal effects. Ables, A. Z., & Nagubilli, R. (2010). Prevention, recognition, and management of serotonin syndrome. *American family physician*, *81*(9), 1139–1142.

[38]For examples see e.g. the section on Compassion Training in Chap. 2.

exercises. Indeed, what is often called 'wearables' is far from an extreme preference and while many devices still are fairly pricey it is becoming increasingly common. In a recent analysis of the market development for wearables research institute IDC forecasts that total global wearables sales will exceed 19 million units by the end of 2014 and in another four years international sales will hit 111.9 million units (i.e. a 500 % increase).[39]

For a good example, consider the success of the devices launched by Apple.[40] At the time of writing the most recent one includes a miniaturized system-on-a-chip which can store a large range of sensors like sweat detectors and motion sensors. Notably their development team included experts in medical sensors in addition to the more traditional engineers and designers—perhaps an indication that wearables are turning into more than 'fashion statements' or 'fitness gadgets'. The aspiration might well be to create a device that deliver information and assist in maintaining, as well as developing, mental and physical health and well-being.

An example of wearables issued by the professional healthcare side (as opposed to the industry) which are intended to treat people who suffer from some form of health-impairment (as opposed to *preventing* ill health as in the Apple case) would be EEG headsets. These EEG systems are designed for at home cognitive training, most often for people who suffer high levels of anxiety or who need attention training. While there has been a large increase in the last two years, the beneficial effects are not clear and there are massive differences in quality between the producers. Key problems include poor signal quality and optimal placement of electrodes. In addition, a host of ethical aspects attach to the project of prescribing, or recommending, EEG devices for cognitive training in the home.[41]

A much more extreme form of enhancement through technology comes in the way of invasive brain-computer interfaces (BCIs).[42] Here the technology is integrated in the body (the brain) and connects the nervous system to a device which enables information exchange. The brain signals are recorded by a computer system which then can be used to control various functions and/or enable the subject to do things they otherwise could not.[43] Examples range from cochlear implants, to deep

---

[39]Source: IDC, Worldwide Wearable Computing Device 2014–2018 Forecast and Analysis, published in March 2014.

[40]http://9to5mac.com/2014/09/06/apple-wearable-to-run-third-party-apps-big-developers-already-seeded-sdk/.

[41]For an example of interesting research see Neuroethics Research Unit, Institut de recherches cliniques de Montreal (IRCM), and Department of Neurology and Neurosurgery, McGill University.

[42]BCIs can be invasive e.g. artificial vision and *motor neuroprosthetics* and non-invasive e.g. EEG.

[43]For a review see e.g. Wodlinger, B., Downey, J. E., Tyler-Kabara, E. C., Schwartz, A. B., Boninger, M. L., & Collinger, J. L. (2015). Ten-dimensional anthropomorphic arm control in a human brain—machine interface: difficulties, solutions, and limitations. *Journal of neural engineering*, *12*(1), 016011.

brain stimulation to stop tremor in Parkinson patients—both techniques are fairly common today—to more distant possibilities such as nerve stimulators in the spine or the brain which would enable people who are paralyzed by spinal cord injuries to move again.[44] This field is broadly labelled 'neuro-prosthetics' and research has been going on since the early 70s.[45]

Traditionally the scientists have focused on developing devices that allow us to bypass neural deficits and augment functions but more recently there have also been advances as to the actual restoring of functions in patients with sensory and/or motor disabilities.[46] An example would be the research into so called bi-directional brain-computer interfaces which enable restoration of tactile sensations in a prosthesis.[47] To restore touch is of course key to motor control for the individual (e.g. it enables them to feel the touch of another human being and it becomes possible for them to pick up delicate objects without crushing them) but it has also turned out that when a sense of touch is restored the patient experiences a strong sense of ownership of the artificial limb.[48] Even more cutting edge is the research which seeks to restore a sense of proprioception (i.e. a sense of where in space their artificial limb is without actually looking at it) to this patient group.[49]

---

[44]Although very interesting the research is still experimental and high risk and many ethical aspects attach. Collinger, J. L., Kryger, M. A., Barbara, R., Betler, T., Bowsher, K., Brown, E. H., ... & Boninger, M. L. (2014). Collaborative Approach in the Development of High-Performance Brain–Computer Interfaces for a Neuroprosthetic Arm: Translation from Animal Models to Human Control. *Clinical and translational science*, *7*(1), 52–59. Iuculano, T., & Kadosh, R. C. (2013). The mental cost of cognitive enhancement. *The Journal of Neuroscience*, *33*(10), 4482–4486. Kadosh, R. C., Levy, N., O'Shea, J., Shea, N., & Savulescu, J. (2012). The neuroethics of non-invasive brain stimulation. *Current Biology*, *22*(4), R108–R111.

[45]Vidal, J. J. (1973). Toward direct brain-computer communication. *Annual review of Biophysics and Bioengineering*, *2*(1), 157–180.

[46]E.C. Leuthardt et al., "Using the electrocorticographic speech network to control a brain-computer interface in humans," *J Neural Eng*, 8:036004, 2011. W. Wang et al., "An electrocorticographic brain interface in an individual with tetraplegia," *PLOS ONE*, 8:e55344, 2013.

[47]In human subject see e.g. Fishel, J. A., & Loeb, G. E. (2012). Bayesian exploration for intelligent identification of textures. *Frontiers in neurorobotics*, *6*. Su, Z., Fishel, J. A., Yamamoto, T., & Loeb, G. E. (2012). Use of tactile feedback to control exploratory movements to characterize object compliance. *Frontiers in neurorobotics*, *6*. In primates see e.g. O'Doherty, J. E., Lebedev, M. A., Ifft, P. J., Zhuang, K. Z., Shokur, S., Bleuler, H., & Nicolelis, M. A. (2011). Active tactile exploration using a brain-machine-brain interface. *Nature*, *479*(7372), 228–231.

[48]Marasco, P. D., Kim, K., Colgate, J. E., Peshkin, M. A., & Kuiken, T. A. (2011). Robotic touch shifts perception of embodiment to a prosthesis in targeted reinnervation amputees. *Brain*, *134*(3), 747–758.

[49]See e.g. The Hopkins Revolutionizing Prosthetics http://www.jhuapl.edu/prosthetics/.

Although very interesting and promising the research is still experimental and high risk and many ethical aspects attach.[50]

A promising alternative to neuro-surgery and deep-brain stimulation might be optogenetics. Here the idea is the neurons in the brain can be manipulated and controlled through light (i.e. not only by electrical impulses as in the above examples). The research is experimental and neuromodulation is currently only performed on animals (e.g. mouse brain tissue). Very briefly the idea is that by genetically modifying opcines in the brain one can control the light sensitive ion channels, which, in turn, gives a chance to open and close a number or cells in the brain, then larger network functioning, and subsequently behaviour. It is hoped that this can be a way to treat patients with severe epilepsy,[51] and schizophrenia. Similarly to the BMI and BCI techniques this holds great promise but also involve extreme risks. Today such medical technologies are still a long way away from human trials and even further from clinical treatments.

It might be worth noticing that the lines between playing computer games in the traditional way (e.g. through a game console or an online set-up) and non-invasive brain stimulators is becoming blurred. Consider for example the recently launched headset designed to improve normal people's capacity for attention while playing a game (through electrodes placed on the forehead)[52] or the gaming glasses promising a virtual reality experience by giving the gamer all kinds of stimulus input.[53] While there might be ample room for improvement on the technology front it does not seem overly speculative to assume that we will see more and more of this type of technology integrated in society. Neuro-prosthetic devices might well become a part of our daily lives and something which we take for granted.

3.6 Possible Effects of Meditation Techniques and Mental Training

Much has been written on life-style choices involving regular physical exercise and certain type of diets (e.g. cardio-vascular training and a diet rich in omega acids) and how this impact our ability to function well both physically and mentally.

---

[50] Collinger, J. L., Kryger, M. A., Barbara, R., Betler, T., Bowsher, K., Brown, E. H., … & Boninger, M. L. (2014). Collaborative Approach in the Development of High-Performance Brain–Computer Interfaces for a Neuroprosthetic Arm: Translation from Animal Models to Human Control. *Clinical and translational science*, *7*(1), 52–59. Iuculano, T., & Kadosh, R. C. (2013). The mental cost of cognitive enhancement. *The Journal of Neuroscience*, *33*(10), 4482–4486. Kadosh, R. C., Levy, N., O'Shea, J., Shea, N., & Savulescu, J. (2012). The neuroethics of non-invasive brain stimulation. *Current Biology*, *22*(4), R108–R111.

[51] Kokaia, M., & Ledri, M. (2013). An optogenetic approach in epilepsy. *Neuropharmacology*, *69*, 89–95.

[52] See for example http://www.foc.us/, this would be the commercial version of the EEG headsets developed for medical purposes mentioned above, and http://www.oculus.com/rift/.

[53] There are also whole body suits for gamers on the market http://priovr.com/.

As previously explained, we agree that an efficient way of improving our cognitive skills, at the present time, is through life-style changes but we have chosen to focus on the changes that regular mental training in the form of meditation can have on the adult brain. We have opted for this instead of more traditional examples as diet, supplements and cardiovascular training as we consider such methods to be almost universally recognized and to discuss them here would consequently add little novelty to our argument.[54]

From a Western neuroscience perspective the study of the therapeutic effects of meditation is relatively new field. In the last 10 years or so interest has, however, grown rapidly (a contributing reason would of course be the availability of technology required to measure effects such as imaging methods e.g. CT and MRI and functional methods, e.g. EEG, fMRI and the integration of different methods). Numerous evidence based studies of high quality have shown that the adult brain is plastic enough to perform structural and functional changes. Some such changes can be brought about through practices where meditation is a core component. Indeed, some of the results appear to corroborate the more anecdotal Eastern view of the therapeutic effects of meditation.

In Chap. 2 we discussed the effects of (i) focused attention training, (ii) open monitoring and mindfulness training and (iii) emotional regulation and compassion training. We wrote about what such (and other techniques of a similar kind) might achieve. We also touched upon some potential risks and the challenges of how to measure and verify the impact. Notably we mainly restricted our references regarding biological effects of the practices to the structural and functional neuroimaging studies, which now enhance our understanding of the neural processes associated with meditation.[55] In Chap. 4 we will take a closer look at how such changes can be cashed out as improved cognitive skills (e.g. cognitive flexibility). Already here, however, we would like to mention a couple of hypotheses of what these (and other) mental training techniques might achieve.

---

[54]For links between cognitive capacity and glucose, creatine and amino acids (for example) see Fox, P.T., Raichle, M.E. et al. (1988). Nonoxidative glucose consumption during focal physiologic neural activity. *Science,* 241(4864), 462–464; Rae, C., Digney, A.L. et al. (2003). Oral creatine monohydrate supplementation improves brain performance: a double-blind, placebo-controlled, cross-over trial. *Proceedings of the Royal Society of London Series B, Biological Sciences,* 270 (1529), 2147–2150; Lieberman, H.R. (2003). Nutrition, brain function and cognitive performance. *Appetite,* 40(3), 245–54. Luchtman, D. W., & Song, C. (2013). Cognitive enhancement by omega-3 fatty acids from child-hood to old age: findings from animal and clinical studies. *Neuropharmacology, 64,* 550–565. Tolppanen, A. M., Solomon, A., Kulmala, J., Kåreholt, I., Ngandu, T., Rusanen, M., … & Kivipelto, M. (2014). Leisure-time physical activity from mid-to late life, body mass index, and risk of dementia. *Alzheimer's & Dementia.* Draganski et al. (2004) Neuroplasticity: Changes in grey matter induced by training, *Nature* 427, 311–312.

[55]Marchand, W. R. (2014). Neural mechanisms of mindfulness and meditation: Evidence from neuroimaging studies. *World journal of radiology, 6*(7), 471.

One hypothesis is that meditation can help us to notice both inappropriate "barriers" such as experiential avoidance and various bias which plausibly are mirrored by functional neurophysiological hindrances between different areas in the brain.[56] Becoming better placed to notice such things might well, again by using meditation techniques, increase or change communication between the different parts of the brain. Further, misinterpretations or connections between thought and emotions which affect us negatively might benefit from such attention and cognitive and/or emotional defusion.[57]

More speculatively, mental training and meditation techniques might even facilitate increased connectivity between parts that 'already knew each other' but also enable parts that have been communicating minimally previously to connect more regularly. It does not seem too far-fetched that such changes might translate to a capacity to view things differently—including both ourselves and others and what we 'attach importance to'.

3.7 Computer Games

Evidently, mental training can take many forms and it seems highly likely that the best way forward is to combine various techniques. One very interesting method to hone and cultivate cognitive skills comes in the form of specially designed computer games. Positively, from our perspective, it seems possible to combine the mental training techniques presented here and playing some such games. At the very least it is not the case that such methods need to cannibalize on each other. In addition to personal interest, a contributing reason for selecting this method is that there is relatively robust scientific evidence (large enough cohorts, have been studied for quite some time, not only 1st person experience accounts etc.) as to the effects.

In order to show how these practices might be complimentary we will sketch a brief account of how playing specially designed computer games (for example involving memory training, moral decision-making, attention and focus) could be highly conducive to strengthening a set of core cognitive and practical skills. There are also examples of computer games which can help decrease experiential avoidance (online CBT) and yet others developed to cultivate empathy and (positive) detachment.

---

[56]We define a bias here as a form of heuristic or shortcut that the human brain is prone to when engaging in e.g. decision-making, general assessment of events, ranking how important events/facts are and what to pay attention to in a situation. Unfortunately the agent tends to be unaware of the nature and magnitude of such bias.

[57]Hayes, Steven C.; Kirk D. Strosahl; Kelly G. Wilson (2003). Acceptance and Commitment Therapy: An Experiential Approach to Behavior Change. The Guilford Press.

3.7.1 The Potential Effects of Computer Games[58]

Traditionally, spending time in front of the computer has been said to impede on other more worthwhile activities (e.g. building relationships with people in the real world, being outdoors, reading books and so on) without fully compensating for the benefits that such and other past-times would bring. Indeed, the computer game industry has been accused of causing everything from anti-social behavior to obesity in the young as well as in adults.[59]

More recently, however, it has been shown that playing certain kinds of computer games (of the type strategy games as opposed to ones that mainly train perception and reaction) also can have positive effects. For example it can improve working memory, the capacity for problem-solving, divergent thinking, attention and ability to focus.[60] It is thought that instant feed-back when a mistake is made is very helpful for maintaining focus on the task. For this positive impact on the cognitive capacities to manifest itself, however, we need to play the games frequently, for longer periods of time and at regular intervals/in a structured manner.

When it comes to cognitive capacities, working memory (and its limitations) is attracting a lot of interest. Most adults one can keep approximately 7 things in the working memory and children can keep 4. Long-term memory, on the other hand, seems to be if not unlimited at least able to store vast amounts of information. Hence, working memory is a bit of a bottle neck in the brain and it would plausibly be advantageous to be able to increase it. Both external stress (environment) and internal (e.g. as a result of lack of emotional regulation, insufficient focus or uncontrolled mind-wandering) have a well-documented and very negative impact on working memory,[61] and recently interesting studies on "math anxiety" show that.

[58]A version of the following sections has been previously published in Fröding, B., & Peterson, M. (2013). Why computer games can be essential for human flourishing. *Journal of Information, Communication and Ethics in Society*, *11*(2), 81–91. I would like to thank Martin Peterson for allowing me to use this material here.

[59]See e.g. De Decker, E., De Craemer, M., De Bourdeaudhuij, I., Wijndaele, K., Duvinage, K., Koletzko, B., … & Cardon, G. (2012). Influencing factors of screen time in preschool children: an exploration of parents' perceptions through focus groups in six European countries. *Obesity Reviews*, *13*(s1), 75–84. Sicart, M. (2009). *The ethics of computer games*. MIT Press. Spence, E. H. (2012). Virtual Rape, Real Dignity: Meta-Ethics for Virtual Worlds. *The Philosophy of Computer Games*, 125–142.

[60]For a detailed argument please see Fröding, B., & Peterson, M. (2013). Why computer games can be essential for human flourishing. *Journal of Information, Communication and Ethics in Society*, *11*(2), 81–91. As will be argued in Chapter 4 such cognitive improvements especially with regards to skills, could be conducive to the overall cognitive flexibility of the person. .

[61]Ashcraft, M. H., & Krause, J. A. (2007). Working memory, math performance, and math anxiety. *Psychonomic bulletin & review*, *14*(2), 243–248.

To get an idea of how a new breed of tailored computer games can help us to improve our cognitive capacities and/or knowledge in ways that influence our quality of life consider the following three examples.[62]

1. Games to help managing negative emotions like stress, anger and phobias. A good example is the game Elude (developed at MIT) which is designed to help the families of patients suffering from bi-polar disorder or depression to better understand and read the mental state of their loved one and to (further) develop their sense of empathy.[63]
2. Games for learning subject specific skills and 'life-skills' as broadly conceived. The idea is that educational games can be used to improve life skills (as broadly conceived of) something which can enable people to fare better in life.[64] Consider, for example, an improved capacity to adjust to changes in one's socio-economic environment—to become more adaptable and thus versatile on the labour-market for example. Concrete examples could be that one is better at identifying opportunities, better at judging one's own capacity for things, suffer less misplaced fears, better at assessing risk, better at epistemic deference. Interestingly, some independent game designers/developers show increasing focus on aspects such as the human being and human interaction in a broader and sometimes rather everyday sense. Plausibly, such games and the roles they can play in the pursuit of the good-life and their effects on wellbeing in the wider sense ought to be studied and taken seriously.
3. Games to improve cognitive capacities and decision-making. Some computer games and other online activities help improve our cognitive capacities and make us better at making decisions. Here we are thinking about gamified decision support systems and games that help to improve memory,[65] or games that un-veil our bias and helps to make us more objective.
4. Games that combine the strategies above, such as 1 and 3, e.g. bio-feedback in order to decrease disabling emotions (e.g. math anxiety) and thereby increase cognitive performance.[66]

To be clear then, there is evidence that certain types of computer games are efficient for cultivating and improving some skills and capacities (practical and

[62]For the broader discussion on the social and cognitive dimensions in computational neuroscience see e.g. Dunne, S., & O'Doherty, J. P. (2013). Insights from the application of computational neuroimaging to social neuroscience. *Current opinion in neurobiology*, 23(3), 387–392.

[63]http://gambit.mit.edu/loadgame/elude.php.

[64]This could be done using 3D touch games and 3D add-on, i.e. 'haptic feedback'.

[65]E.g. http://www.lumosity.com.

[66]Verkijika, S. F., & De Wet, L. (2015). Using a brain-computer interface (BCI) in reducing math anxiety: Evidence from South Africa. *Computers & Education*, *81*, 113–122. Learning mathematics is highly associated with attitudes towards mathematics and emotions like math anxiety. This condition is quite common, and by using a BCI solution math anxiety could successfully be reduced.

cognitive).[67] In some cases, computer games could even be *more* efficient than traditional forms of education. That said, there is also a great number of so called 'brain games' (designed by cognitive scientists and neuroscientist) available on the market. Many of these games claim to e.g. train cognitive abilities which are generalizable outside the specific situation and to counter age and disease related neurological degeneration. Such claims need to be substantiated by much more research to be considered solid.[68]

3.8 Combining Methods

As indicate above, this type of specialised computer games have great potential for improving and training core cognitive capacities in the normally functioning brain.[69, 70]

Plausibly, playing such computer games can enable us to train faster and in a more specialized and tailored way than we otherwise would. Additional argument in favor of exploring computer games and online activities in particular is the fact that in the modern society being computer and technology savvy (in the broadest conceivable way) is indeed a necessary prerequisite for accessing information and actively partaking in society. Further to this, there is also the increasing 'gamification', i.e. the practice of including computer game technology in mundane everyday tasks like washing up or exercising so that they become more exciting, in many Western societies. Introducing the element of competition and instant, and rather concrete, feedback is of course very appealing to most people.[71] We can imagine using a range of web-based tools such as computer games, online reporting, chat rooms etc. both in order to increase the motivation, compliance and

---

[67]Søraker, J. H., Van der Rijt, J. W., de Boer, J., Wong, P. H., & Brey, P. (2014). Well-Being in Contemporary Society. Especially chapter 8 'Increasing societal well-being through enhanced empathy using computer games', Annett J., and Berglund S. Granic, I., Lobel, A., & Engels, R. C. (2014). The benefits of playing video games. *American Psychologist, 69*(1), 66.

[68]Note that there is no consensus on the potential for cognitive improvements or slowing down of cognitive degeneration and that much more research is required to confirm the positive effects of so called 'brain games'. See e.g. Statement issued on Oct 14, 2014 by The Stanford Center for Longevity and the Max Planck Institute for Human Development.

[69]Bergman-Nutley, S., & Klingberg, T. (2014). Effect of working memory training on working memory, arithmetic and following instructions. *Psychological research, 78*(6), 869–877.

[70]There are also studies of the effects on the diseased brain see e.g. Zimmermann, R., Gschwandtner, U., Benz, N., Hatz, F., Schindler, C., Taub, E., & Fuhr, P. (2014). Cognitive training in Parkinson disease Cognition-specific vs nonspecific computer training. *Neurology, 82* (14), 1219–1226.

[71]For a good account of how gamification has become an integrated part of society see for example 'Reality is broken: Why Games Makes us Better and how they can Change the World'. The book paints a bright future for gamification. In fact, she argues that it might help to save the world McGonigal, J. (2011). *Reality is broken: Why games make us better and how they can change the world.* Penguin.

commitment as well as improving the actual cognitive capacities. For some concrete examples, consider the type of 'reminder' or 'coaching apps' that facilitate mindfulness training which are available today.[72] It appears likely that computer games which are designed to aid and boost the effects of traditional meditation will be developed.

On the other hand, it is not clear that the type of computer games we have today can help us achieve something radically different than more classical education i.e. that it would prove more than 'just another training method'. Other issues include: lack of robust scientific evidence that these types of computer games can *prevent* cognitive impairment or degeneration; the generalizability of the skills is debatable; it is not clear that physical exercise is less efficient for maintaining and promoting general and cognitive health. As so often more research is needed.

Without ruling out the future potential of alternative methods (such as pharmaceutical drugs, hormones, neurotransmitters, wearable technology and BMI), our position is that given the pressing nature of the problem coupled with current scientific knowledge (i.e. limited), our best bet for lasting cognitive enhancement will come through 'life-style' broadly conceived of. In this we include, for example, meditation, computer games, other relevant online activities, traditional education, physical exercise, diet and supplements.[73]

The effects of cardiovascular training, diet and supplements are well-researched and have not been explored here. Designing computer games to promote cognitive skills is a relatively new field but it seems clear that specialized computer games could work well complementary tool for improving cognitive skills.[74] While the effects are positive they are not strong enough to solve our current cognitive predicament but could work very well in combination with other methods such as focused attention training, open awareness and mindfulness training and emotional regulation and compassion training. The combination of imaging, neurophysiological monitoring and/or treatment modalities such as neuroenhancers, meditation practices and neurofeedback, e.g. "multimodal neuroenhancement," could be more powerful than the use of single methods alone. An especially interesting combination is physical exercise with concomitant mental training.[75] Employing multiple technologies together may add to the expense, time and complexity of studies and

[72]Boettcher J, Aström V, Påhlsson D, Schenström O, Andersson G, Carlbring P. (2014). Internet-based mindfulness treatment for anxiety disorders: a randomized controlled trial. *Behav Ther.* 45(2):241–253. Ly, K. H., Trüschel, A., Jarl, L., Magnusson, S., Windahl, T., Johansson, R., … & Andersson, G. (2014). Behavioural activation versus mindfulness-based guided self-help treatment administered through a smartphone application: a randomised controlled trial. *BMJ open*, *4*(1), e003440.

[73]Evidently much more empirical research as to the effects, the generalizability and how long they last, is required.

[74]See e.g. Clark, V. P., & Parasuraman, R. (2014). Neuroenhancement: enhancing brain and mind in health and in disease. *Neuroimage,* 85, 889–894.

[75]Curlik, D. M., & Shors, T. J. (2013). Training your brain: Do mental and physical (MAP) training enhance cognition through the process of neurogenesis in the hippocampus?. *Neuropharmacology*, *64*, 506–514.

treatment protocols. However, if the results of cost–benefit analyses show that the added benefits outweigh the costs, then multimodal neuroenhancement may be preferred.[76] Indeed, as we will return to in Chap. 6, it is perfectly possible and perhaps even wise to engage in several methods on a regular basis given the urgent need for improvement.

In the next chapter we will; (a) define cognitive flexibility, (b) explain how the plastic effects of the meditation techniques promoted here (see Chap. 2) can increase cognitive flexibility, (c) sketch an argument for how an increased propensity for cognitive flexibility would be conducive to the installing of a set of key epistemic virtues.

---

[76]Clark, V. P., & Parasuraman, R. (2014). Neuroenhancement: enhancing brain and mind in health and in disease. *Neuroimage*, *85*, 889–894.

Chapter 4
Cognitive Flexibility

Abstract This chapter takes a closer look at cognitive flexibility. Firstly, we define this core cognitive capacity and explain why it is good to have it to a high degree. Secondly, we examine the link between the meditation techniques promoted here (see Chap. 2) and increased cognitive flexibility, as well as the impact on other psychological capacities. Thirdly, we point out that high cognitive flexibility (as an example of such improved psychological capacities) does not guarantee responsible moral decision-making. Consequently we need a robust, and action guiding, moral framework which can anchor these capacities and guide vacillating agents. The chapter finishes with a brief discussion of the connection between improved core cognitive capacities and the installing of a set of key epistemic virtues. This subject is then expanded on in Chap. 5.

Keywords Experiential avoidance · Cognitive flexibility · Decision-making · Epistemic virtues · Well-being

4.1 How Does Cognitive Flexibility Relate to Meditation?

In the last three chapters we have explained *that* regular mental training and meditation (Focused Attention Training, Open monitoring and mindfulness training and Compassion Training) can cause lasting biological changes in the brain. Further, we have provided a number of examples and studies of *how* such functional and structural changes can come about.

Now we will explore how some such effects on our biological make-up can improve a set of core cognitive capacities in a lasting and generalizable sense. Our examples include; the ability to focus and shut out stimuli, "controlled" mind-wandering, and improved capacity for emotional regulation (presumably applicable both to oneself and to others). Heightened core capacities of this kind would plausibly promote general emotional stability, reduce stress and misplaced fear,

B. Fröding and W. Osika, *Neuroenhancement: How Mental Training and Meditation Can Promote Epistemic Virtue*, SpringerBriefs in Ethics,
DOI 10.1007/978-3-319-23517-2_4

promote open-mindedness and (possibly) decrease experiential avoidance all of which could contribute to increased general cognitive flexibility as broadly defined.

In Chaps. 5 and 6 we will make the case that some such changes could be connected to an improved capacity for instilling and maintaining a range of character traits (primarily epistemic virtues)[1] which might enable better (as in more responsible and less biased) decision-making. On the virtue ethics account, an individual who has managed to instill all the virtues to a (sufficient) degree would have a better understanding of both moral and practical situations and well placed to make ''all things considered' decisions. It appears that being more cognitively flexible could bring about a host of 'generally improved life-skills' e.g. being more adaptable, better at risk assessment and more impartial. Such and other skills could be manifested in our behavior and the choices we make.[2] This argument (including some comments on the, sometimes complicated, step from 'knowing' to 'doing') will be developed in Chap. 5.

4.2 Cognitive Flexibility

What is cognitive flexibility and why it is good to have it to a greater extent? In practice, to be more cognitively flexible means having an improved capacity for noticing several different concepts simultaneously, and to shift between these sometimes mutually contradictory concepts. It is often used as an umbrella term but a tentative definition of cognitive flexibility is that it "… includes the ability to represent knowledge from different conceptual and case perspectives and then, when the knowledge must later be used, the ability to construct from those different conceptual and case representations a knowledge ensemble tailored to the needs of the understanding or problem-solving situation at hand."[3]

Perhaps cognitive flexibility can be a cornerstone of a more general capacity for visionary combinations and, further, that the fact that we can (and dare) think of new ways of doing things more easily is rather likely to contribute to making our lives go better in an increasingly complex world.

[1]As discussed by Aristotle and some modern virtue epistemologist (e.g. Heather Battaly, John Greco, Ernest Sosa and Linda Zagzebski).

[2]How such good decision-making and pro-social behavior can be promoted by social embedding structures see Chap. 6.

[3]Spiro, R. J., Feltovich, P. J., Jacobson, M. J., & Coulson, R. L. (1996). Cognitive flexibility, constructivism, and hypertext: Random access instruction for advanced knowledge acquisition in ill-structured domains. I Simpósio Investigação e Desenvolvimento de Software Educativo. An earlier version of this chapter originally appeared in two parts in the journal Educational Technology (1991, 11(5), 24–33 and 1991, 11(7), 22–26). p. 24.

Cognitive flexibility is associated with a range of positive abilities e.g. capacity to focus, capacity to suppress irrelevant stimuli, capacity to control one's thinking better (for more on controlled mind-wandering see Chap. 2) and not get caught in suboptimal, rigid thought habits, which presumably is promoted by an increased ability to connect (or disconnect) various parts of the brain and alter their communication.

Factors that undermine cognitive flexibility include stress and cognitive dissonance. Both states are (to an extent) the result of a low capacity for supressing irrelevant stimuli and poor emotional regulation. Low levels of such cognitive skills (may) result in poor meta-awareness and a tendency to experiential avoidance. Experiential avoidance is a putative dysfunctional process recognized by a number of theoretical orientations. Experiential avoidance has been described as the phenomenon that occurs when a person is unwilling to remain in contact with particular private experiences (e.g., bodily sensations, emotions, thoughts, memories, behavioral predispositions) and takes steps to alter the form or frequency of these events and the contexts that occasion them.[4]

Meditation practices as well as compassion training could decrease the tendency to experiential avoidance, and increase cognitive flexibility. In a recent study More and Malinowski investigated the link between meditation, self-reported mindfulness and cognitive flexibility as well as other attentional functions. They compared a group of meditators experienced in mindfulness meditation with a meditation-naïve control group on stress measures and on concentration and endurance tests. Overall the results showed that attentional performance and cognitive flexibility are positively related to meditation practice and levels of mindfulness. This suggests that mindfulness is intimately linked to improvements of attentional functions and cognitive flexibility, potentially very important for mental balance and well-being.[5] This is also a central area of research in e.g. acceptance and commitment therapy and relational frame theory.[6] These improved cognitive capacities could be described as a stepping stone between 'the structural and functional changes in the brain' and 'capacity to instil the epistemic virtues'. It fits well with the virtue ethics idea that the combination of the virtues and careful deliberation leads the agent to act. Plausibly heightened awareness in the sense described above would be conducive to the deliberation.

---

[4]Barnes-Holmes, Y., Hayes, S. C., Barnes-Holmes, D., & Roche, B. (2002). Relational frame theory: A post-Skinnerian account of human language and cognition. *Advances in child development and behavior, 28*, 101–138. Torneke, N. (2010). *Learning RFT: An introduction to relational frame theory and its clinical application*. New Harbinger Publications.

[5]Moore A, Malinowski P. Meditation, mindfulness and cognitive flexibility. *Conscious Cogn*. 2009 Mar;18(1):176–86.

[6]Hayes, S. C., Strosahl, K. D., & Wilson, K. G. (1999). *Acceptance and commitment therapy: An experiential approach to behavior change*. Guilford Press.

4.3 Why Is It Good to Be More Cognitively Flexible?

A heightened capacity for cognitive flexibility (as well as meta awareness and improved emotional regulation) would plausibly bring with it a set of improved life-skills (mentioned previously in Chaps. 1 and 3) e.g. adaptability to novel circumstance, more rationally placed fears, more accurate risk-assessment, impartiality and tolerance. Such and other similar capacities would serve anyone well in life regardless of their individual life-plan and broader idea of what constitutes the good life. If that is accepted this could be taken as an argument in support for actively encouraging people to engage in these types of meditation (for a discussion see Chap. 6).

It might even be thought that if there is evidence that mindfulness and meditation training increases e.g. cognitive flexibility[7] and has no negative side-effects, that perhaps even young children should be introduced to such practices.[8] Contrast it with how we nowadays encourage physical exercise in both the young and the old —could this not be extended to mental training? An increase in cognitive flexibility and emotional relation skills is likely to enable us to make better decisions and suffer less cognitive dissonance. In short, it influences how well our lives go and, further, it appears compatible with a pluralistic idea of a right to an open future. It might indeed open more 'doors', as people might become less conflicted.

Not implausibly, some of the observed changes in the brain could be connected to (i) a set of character traits and/or (ii) the capacity to develop and maintain some such traits.

A lot of our decisions are made sub-consciously but during those processes we are none the less capable of some kind of reasoning. In order for that reasoning to lead to beneficial decisions however, it needs to be properly trained and 'groomed'. Such training can come in the form of past experiences, including, for example, meditation and mental training. This is especially useful when time is scarce as it enables the brain to reach a beneficial decision much quicker. One can also work with the outcome of the decisions, and focus on behavioral changes wanted.

We can prime/groom these 'subconscious' decision-making processes into different "directions" as broadly conceived. In other words, while we still cannot fully control the decision-making processes we can improve the chances of the outcome being one which is in line with our valued direction. A concrete way of doing just that is to practice virtuous deliberation in similar scenarios, and also to prepare for such deliberations.[9]

---

[7]Cassidy, S., Roche, B., & Hayes, S. C. (2011). A relational frame training intervention to raise Intelligence Quotients: A pilot study. *The Psychological Record*, *61*(2), 2.

[8]Flook, L., Goldberg, S. B., Pinger, L., & Davidson, R. J. (2015). Promoting prosocial behavior and self-regulatory skills in preschool children through a mindfulness-based kindness curriculum. *Developmental psychology*, *51*(1), 44.

[9]E.g. using some of the methods presented in Sects. 1.5.1 and 2.6 (e.g. focused attention, open monitoring and compassion training).

4.4 Improved Cognitive Flexibility Can Translate to Better Decision-Making

4.4.1 *One Choice Too Many*

Making choices is often difficult and, perhaps counter intuitively, the more information we access and the more options we have the worse it tends to get.[10] This is 'the paradox of choice'—too many options undermine our subjective wellbeing and decision-making competence. We quickly go from well-informed to over-informed and thus less able to make beneficial choices. Related to this is a state that many of us are all too familiar with namely 'decision-paralysis' i.e. when we get so overwhelmed by all the options that we cannot make a choice.[11]

Since it appears unlikely that the information flow will slow down a good strategy for reducing this cognitive dissonance and the bad decisions that (can) follow, is to structure the information differently. Recent research has shown how 'decision-paralysis' or 'choice overload' can be mediated by (i) the structure of the decision problem and (ii) which strategy the person facing the choice deployed.[12] There are many techniques for this and clearly some are more efficient than others. Unfortunately, learning and practicing these models might not be enough. To advance significantly in the choice-making domain we would benefit from improved cognitive capacities. An example would be a heightened ability for deliberating complicated issues (e.g. courtesy of better capacity to focus on the relevant features of the given situation, high meta-awareness and a reduced tendency to experiential avoidance) which might issue in decisions more conducive to general well-being. But to achieve that we would need to change the biological set-up in the brain.

As explained in previous chapters the last few decades have witnessed massive breakthroughs in neuroimaging techniques. The technology has been used to unravel experience dependent plasticity and we have learnt a lot about the mechanisms of plasticity and the relations between brain activity and behaviour. While plastic changes are limited by existing connections (which are the result of genetically controlled neural development and are ultimately different across

[10]For a good introduction to some of the harmful consequences of information overload see e.g. Eppler, M. J., & Mengis, J. (2004). The concept of information overload: A review of literature from organization science, accounting, marketing, MIS, and related disciplines. *The information society*, *20*(5), 325–344. Iyengar, S. S., & Lepper, M. R. (2000). When choice is demotivating: Can one desire too much of a good thing?. *Journal of personality and social psychology*, *79*(6), 995. Bawden, D., & Robinson, L. (2009). The dark side of information: overload, anxiety and other paradoxes and pathologies. *Journal of information science*, *35*(2), 180–191. Hendricks, V. F., & Hansen, P. G. (2014). *Infostorms: How to Take Information Punches and Save Democracy*. Springer Science & Business Media.

[11]Schwartz, B. (2004). The paradox of choice: Why less is more. *New York: Ecco*.

[12]Scheibehenne, B., Greifeneder, R., & Todd, P. M. (2010). Can there ever be too many options? A meta-analytic review of choice overload. *Journal of Consumer Research*, *37*(3), 409–425.

individuals) the *reinforcement of existing connections* is the consequence of environmental influences.[13]

Such environmental factors include (amongst other things) life-style choices. Examples range from diet and cardiovascular exercise to meditation techniques and various forms a mental training (e.g. specialized video or computer games).[14]

As previously explained the effects of mental training techniques (including Focused Attention Training, Open monitoring e.g. mindfulness training and Compassion Training) can be linked to increased cognitive flexibility. All three techniques can have concrete, lasting, structural and functional changes in the human brain, of a kind that plausibly could be taken to have a positive impact on our cognitive capacities.[15] In addition to improved cognitive flexibility and raised meta-awareness documented examples include: improved focus, emotional stability, improved capacity for introspection, reduced stress and fear, more compassionate behavior and improved sense of impartiality.

More speculatively, but not unlikely, such fortified cognitive skills would be generalizable outside the original training scenario. They might enable us both to pick up on, and be sensitive to, the relevant features of the choice-situation i.e. structure and handle the available information better. The lowering of stress in the body could open the door to a reduction of misplaced fear, decreased experiential avoidance and make us more able to balance the effects of various systemic (cognitive) bias. Such changes might, in turn, make it easier to understand and instill a set of epistemic virtues—or more broadly speaking 'life-skills'—which are key to responsible decision-making. In addition, it could increase the capacity for identifying the alternatives. The empowering of the individual is highly compatible with liberal values, pluralism and the modern, democratic society.

As concluded in the last chapter it seems that of the cognitive enhancement techniques which are available today (bearing efficiency and degree of risk in mind) a combination of a range of life-style choices with classical educational and training

---

[13]Draganski, B., Gaser, C., Busch, V., Schuierer, G., Bogdahn, U., & May, A. (2004). Neuroplasticity: changes in grey matter induced by training. *Nature*, *427*(6972), 311–312. Pascual-Leone, A., Amedi, A., Fregni, F., & Merabet, L. B. (2005). The plastic human brain cortex. *Annu. Rev. Neurosci.*, *28*, 377–401. Pinho, A. L., de Manzano, Ö., Fransson, P., Eriksson, H., & Ullén, F. (2014). Connecting to Create: Expertise in Musical Improvisation Is Associated with Increased Functional Connectivity between Premotor and Prefrontal Areas. *The Journal of Neuroscience*, *34*(18), 6156–6163.

[14]Granic, I., Lobel, A., & Engels, R. C. (2014). The benefits of playing video games. *American Psychologist*, *69*(1), 66. Greitemeyer, T., & Osswald, S. (2010). Effects of prosocial video games on prosocial behavior. *Journal of personality and social psychology*, *98*(2), 211. Sestir, M. A., & Bartholow, B. D. (2010). Violent and nonviolent video games produce opposing effects on aggressive and prosocial outcomes. *Journal of Experimental Social Psychology*, *46*(6), 934–942.

[15]Lutz, A., Slagter, H. A., Dunne, J. D., & Davidson, R. J. (2008). Attention regulation and monitoring in meditation. *Trends in cognitive sciences*, *12*(4), 163–169. Luders, E., Clark, K., Narr, K. L., & Toga, A. W. (2011). Enhanced brain connectivity in long-term meditation practitioners. *Neuroimage*, *57*(4), 1308-1316. Ricard, M., Lutz, A., & Davidson, R. J. (2014). Mind of the Meditator. *Scientific American*, *311*(5), 38-45. Moore A, Malinowski P. Meditation, mindfulness and cognitive flexibility. Conscious Cogn. 2009 Mar;18(1):176–86.

would be a good option. Consequently, it is not an either/or-situation. In fact, it is rather plausible that the various tactics might prove complementary as opposed to rivalrous.[16]

4.4.2 *Cognitive Bias*

Broadly speaking a bias is a form of heuristic or shortcut that the human brain is prone to when engaging in e.g. decision-making, general assessment of events, ranking how important events/facts are and what to pay attention to in a situation.[17]

Human reasoning is subject to many bias and we are in many cases unaware of how they affect our lives and might undermine good decision-making. With great simplification we can split them into two groups. Some bias are the results of our biological make-up and other are the results of social conditioning (e.g. taught through culture, tradition and religion) or circumstance (e.g. the type of bias that might afflict a patient partaking in a non-blinded study). Furthermore, biases (as so many other human characteristics) are results of the interplay between nature and nurture.[18]

Biologically caused bias would, plausibly, have its origins in what would have been an evolutionary advantage. Individuals with such and such bias would have had a tendency towards choices and behavior which would have been (more) advantageous to their survival (fitness criteria). As evolution is a very slow process (which admittedly is a good thing on occasion) we are still acting on some of those bias even though they have ceased to be advantageous a long time ago. Indeed, some of them frequently leads us wrong and contribute to poor decision-making. One such false friend is confirmation bias. On a very general level confirmation bias is the tendency to treat (and even notice) facts that confirms already existing beliefs more favorably than facts that are either neutral or contradicts the beliefs already held. It affects how we gather, interpret and recall information.[19]

---

[16]Fröding, B. E. E. (2011). Cognitive enhancement, virtue ethics and the good life. *Neuroethics, 4* (3), 223–234.

[17]For a good overview see Gilovich, T., Griffin, D., & Kahneman, D. (Eds.). (2002). *Heuristics and biases: The psychology of intuitive judgment*. Cambridge University Press.

[18]Rutter, M. (2006). *Genes and behavior: Nature-nurture interplay explained.* Blackwell Publishing.

[19]Devine, P. G., Hirt, E. R., & Gehrke, E. M. (1990). Diagnostic and confirmation strategies in trait hypothesis testing. *Journal of Personality and Social Psychology*, *58*(6), 952. Nickerson, R. S. (1998). Confirmation bias: A ubiquitous phenomenon in many guises. *Review of general psychology*, *2*(2), 175. Baron, Jonathan (2000), *Thinking and deciding* (3rd ed.), New York: Cambridge University Press.

Evidently, this type of cognitive bias can have a detrimental effect of the quality of our decision-making and lead to over-confidence. Unfortunately, it is hard to detect bias in oneself and even harder to control.[20]

Another example which frequently trips up decision-making is the status quo bias.[21] This is a cognitive bias and might be broadly described as preferring an option to another in virtue of it preserving the status quo. Since the preference is not based on the status quo *actually being better (as in preferable)* in an objective or subjective sense, it is hard to regard this as a rational preference.[22] The risk is that status quo bias undermines our judgment and decision-making by prompting us to reject things simply in virtue of them being novel or for deviating from what we perceive as normal.[23]

From the perspective of this book, the status quo bias might be especially unfortunate as it can put a spanner in the works of the positive spiral that we are sketching.[24] The idea is that the effects of life-style changes on cognitive capacities would make us better at judging what the right decision might be but also—and perhaps even more importantly—more likely to keep up the good habits and thus create a self-reinforcing cycle. Clearly our 'enhancement model' is not a one off adjustment or treatment—this is a commitment to a set of values and the virtuous life.

In summary then; when the human brain strives to reach coherence and thus lower the stress-level, motivated reasoning[25] in combination with a set of other bias (e.g. status quo bias and confirmation bias) comes into play. Unsurprisingly, this has a rather negative impact on our decision-making. In this chapter we have built the case that improved core cognitive capacities like cognitive flexibility and meta-awareness are likely to increase the ability to notice our own biases and possibly reduce motivated reasoning. As will be shown over the next chapters this can be conducive to the development of a set of epistemic virtues. Such virtues

---

[20]Other examples of bias include e.g. loss aversion, sunk cost, lacking capacity for risk assessment, over optimism, framing effects, substitution (i.e. replacing the tricky problem at hand with a simple one pretending that they are analogous or at least similar enough). For a good discussion see Kahneman, D. (2011). *Thinking, fast and slow*. Macmillan.

[21]See e.g. Kahneman, D., & Tversky, A. (1984). Choices, values, and frames. *American psychologist*, *39*(4), 341.

[22]For a method for how to potentially reduce the influence of this bias see Bostrom, N., & Ord, T. (2006). The Reversal Test: Eliminating Status Quo Bias in Applied Ethics*. *Ethics*, *116*(4), 656–679.

[23]Bostrom, N., & Roache, R. (2010). Smart policy: cognitive enhancement and the public interest. *Contemporary Readings in Law and Social Justice*, (1), 68–84.

[24]As pointed out by Nick Bostrom: while it does not seem like the status quo bias has or is serving some highly important function one should be open to the possibility that this might be the case. That said, such reservations do not seem to target our 'enhancement model' (as opposed to one off treatments with drugs or technology).

[25]By this we mean decisions based on unreflected gut reactions as opposed to people's strong moral intuitions which would be well considered, stable and able to withstand the test of time.

could compensate for, to various degrees, some of our cognitive shortcomings and, in combination with boosted cognitive skills improve of our decision-making (both the quality and propensity to act in accordance).

4.5 Smarter but Not Nicer

Unfortunately, it is far from clear that improved cognitive capacities would lead people to actually make morally better (e.g. less selfish and more long-term sustainable) decisions. The fact that one might be more cognitively flexible does not mean that one is automatically also better at moral decision-making. Quite to the contrary it is all too easy to envision a cognitively enhanced person using her capacity to manipulate and exploit others.[26]

The skills attained in meditation can (and have) also been used in warfare and power struggles, manipulation of others etc. When the meditative schooling of attention first made its way into the West from Asia, it was used, for example, by the military and security agencies. They quickly recognized that improved *samadhi* or "single-pointed attention" would be highly useful in a combat or surveillance situation. Example from civil society would also include sport teams (e.g. shooting[27] and basketball) who have used the techniques to reduce stress levels and increase focus in their players.[28]

Consequently there is a need for a framework which is likely to influence the agents to make the morally better choice. In the Buddhist tradition this is called *sila* or virtue, and it is held to be the cornerstone of the Noble Eightfold Path. Within this tradition the practices of right speech, right action, and right livelihood are understood as essential to moral development. For those undergoing training within the Buddhist tradition a rather demanding set of ethical precepts or rules have to be observed (five for lay practitioners and 227 for ordained monks).

In a secular Western context, however, the enforcement of many such social rules could be perceived to conflict with liberal values and to undermine, or even violate, autonomy. While moral guidance might, on the one hand, be recognized as helpful there is little consensus with regards to the extent and the source of such advice. This book suggests that virtue ethics is especially well suited to create the moral fabric that can help anchor the cognitive capacities. One reason to favor virtue ethics as a candidate for such an embedding structure is that it, in contrast to

[26]Baron Cohen S. (2011) The Science of Evil: On Empathy and the Origins of Cruelty. Basic Books, New York, NY. Decety, J., Chen, C., Harenski, C. L., & Kiehl, K. A. (2013). An fMRI study of affective perspective taking in individuals with psychopathy: imagining another in pain does not evoke empathy. Frontiers in Human Neuroscience, 7:489.

[27]Solberg, E. E., Berglund, K. A., Engen, O., Ekeberg, O., & Loeb, M. (1996). The effect of meditation on shooting performance. *British journal of sports medicine*, *30*(4), 342–346.

[28]Birrer, D., Röthlin, P., & Morgan, G. (2012). Mindfulness to enhance athletic performance: Theoretical considerations and possible impact mechanisms. *Mindfulness*, *3*(3), 235–246.

some of the more paternalistic alternatives, sits relatively comfortably with the idea of liberal democracy that many subscribe to. On the virtue account what is good and rational for the individual (most often) overlaps with that which is pro-social.

4.6 Summary

On a very general level, it appears that meditators achieve an increased capacity to

(i) change the connectivity between various parts of the brain (both increase and decrease), and make them work together/communicate and,
(ii) detach from the self and take a step back

In combination, this improved ability to connect and detach enable the meditators to view things differently. Plausibly a certain detachment from the self and one's immediate interests would result in a type of decision-making which is less of a slave to inbuilt bias. More speculative, but not unlikely, detachment could enable the agent to view both herself, others and general problems (global warming, racism, conflicting views on issues like religion etc.) in a different light. This would imply a fuller understanding of the situation at hand as well as the broader context.

We have also suggested that heightened core capacities plausibly can issue in functional improvements for the individual. It could, for example, promote general emotional stability, reduce stress and misplaced fear, promote open-mindedness and (possibly) decrease experiential avoidance all of which could contribute to increased general cognitive flexibility as broadly defined. In Chap. 5 we will show how such changes could be connected to an improved capacity for instilling and maintaining a range of character traits (primarily epistemic virtues) which might enable better (as in more responsible and less biased) decision-making.

Chapter 5
Some Key Elements of Virtue Ethics

Abstract The purpose of this chapter is two-fold. In Part A we explain some key aspects of virtue ethics e.g. including *eudaimonia* (the good life), the concept of the virtues and the development of stable character traits. For space reasons this will be very brief but should provide some theoretical background for the more general discussion in this book. To be clear, this is not intended as an authoritative, or exegetic, reading of Aristotle. Rather, our aspiration is to suggest that many of the ideas in the Nicomachean Ethics (Aristotle's central work on ethics) make for a highly useful approach to modern moral problems. In Part B we focus on epistemic virtues, both traditional and modern, and provide examples of their role in decision-making. We also show how the cognitive improvements from previous chapters can both increase the commitment to the type of life described by Aristotle, and boost the capacity for cultivation the necessary epistemic virtues.

Keywords Aristotle · Epistemic virtue · Stable traits · Eudaimonia · Character · Mental training · Moral framework

5.1 Part A: A Brief Introduction to Virtue Ethics

5.1.1 Introduction

Chapter 1 introduced the current predicament i.e.—the combination of a rapidly changing society and research confirming that our cognitive capacities are more limited, and our bias more substantive, than what was previously known.[1] To address this great challenge and improve our decision-making skills we suggested that there was a need for cognitive improvements.

[1]We understand bias as a form of heuristic or shortcut that the human brain is prone to when engaging in e.g. decision-making, general assessment of events, ranking how important events/facts are and what to pay attention to in a situation. Unfortunately the agent tends to be unaware of the nature and magnitude of such bias.

B. Fröding and W. Osika, *Neuroenhancement: How Mental Training and Meditation Can Promote Epistemic Virtue*, SpringerBriefs in Ethics,
DOI 10.1007/978-3-319-23517-2_5

In Chap. 2 we described three methods for improving a set of cognitive and emotional regulation skills and gave an account of the neurophysiological background for the development of such capacities. The methods are Focused Attention Training, Open monitoring e.g. in mindfulness training, and Compassion Training. All three have, in evidence based studies, been shown to have lasting and generalizable effects on some core cognitive capacities.

In Chaps. 3 and 4 we introduced four such core cognitive capacities: cognitive flexibility, increased focus/attention (which in turn is related to as meta-awareness), controlled mind-wandering and improved capacity for emotional regulation. Many people would benefit from having these skills to a higher (and more stable) degree as they are strongly correlated with how well our lives go and how successful we are at obtaining (and maintaining) those things that are of objective and subjective value.

For concreteness we chose some examples of which practical skills improvement of these core capacities can issue in. These are functional and behavioral changes which expresses themselves in our daily life. For example, to be more cognitively flexible can translate to suffering less misplaced fear and fewer cognitive bias (both as in being better to compensate for them and/or control them). It could also make us more creative and encourage divergent thinking. Increased meta-awareness in combination with a capacity for emotional regulation can issue in e.g. feeling empathy (and acting compassionate) and being better at detaching from the self and to engage in introspection.

But, as many of us might have experienced—the step from "knowing what to do" and "actually doing it" is not always entirely straight forward. Therefore, the rest of the book is devoted to showing how the suggested cognitive improvements can be connected to the development a set of stable traits (virtues) which would be conducive to good decision-making.

5.1.2 A Suitable Moral Framework[2]

As commonly known, high cognitive flexibility does not guarantee moral behavior and good decision-making (in the impartial, responsible, pro-social way we have in mind). It is easy to imagine that, for example, high trait creativity could correlate with a tendency to being risk prone and thus fail to pay proper attention to prudence and moderation. Or, on a darker note, that the cognitively improved would use their novel skills to manipulate and exploit the people around them. Consequently we need a moral framework which is robust and detailed, yet flexible enough, to provide moral advice (including action guiding) in an increasingly complex world.

[2]Parts of this chapter has been previously published in Björkman, B. (2008). Virtue Ethics, Bioethics, and the Ownership of Biological Material.

This chapter seeks to give some reasons for which virtue ethics is especially well suited to help us with this.[3]

We will argue that improved cognitive skills of the type previously described are likely to be conducive to understanding and embracing the type of life that Aristotle described as the virtuous life. Amongst other things this involves instilling, and acting in accordance with, a set of epistemic and moral virtues. The examples used here are primarily epistemic and include intellectual honesty, intellectual courage, open-mindedness, tolerance, impartiality, commitment to fairness, capacity for introspection and detachment and improved memory (both with regards to process memory and long-term memory) (see Sect. 5.2.8 below).[4]

We have chosen those examples for two reasons. Firstly, these are skills which plausibly could improve and deepen our understanding of what is morally expected of us when we engage in decisions regarding, for example, modern technology (e.g. nano-technology and biotechnology), political strategies and scientific research (including medicine). Having such virtues might enable us to identify and approach some of the information-related problems and risks that we encounter in a more responsible manner than we otherwise would.

Secondly, these virtues would plausibly issue in decisions and behavior which is (more) pro-social and less likely to lead to conflict and crisis. On our understanding 'good' or 'responsible' decision-making would involve a sense of equity and a capacity and willingness to make all things considered judgments. Further, the more (epistemically) virtuous one becomes the deeper the understanding of why the good life (as described by Aristotle) is, all things considered, a better option than the alternatives. This re-enforces the commitment and motivation and creates a positive spin (see Chap. 6).

To see why virtue ethics might be especially suitable when we seek to address a number of ethically challenging issues in decision-making and information handling, the reader needs to be familiar with some of the central ideas in Aristotle's Nicomachean Ethics. As the format of this book hardly allows for much more than scraping the surface of this rich theory we aspire only to give the briefest of introductions.[5]

---

[3]The focus of this book is cognitive enhancement (as opposed to physical and moral enhancements) and how that could be achieved through committing to meditation and virtue ethics. Note however, that some philosophers have argued that in order to handle the potential dangers of cognitive enhancement humans need moral enhancement and, further, that this would be best brought about not through virtue ethics but rather by pharmaceutical drugs and or hormones. For an interesting account see Douglas, T. (2008). Moral enhancement. *Journal of applied philosophy, 25*(3), 228–245. Persson, I., & Savulescu, J. (2008). The perils of cognitive enhancement and the urgent imperative to enhance the moral character of humanity. *Journal of Applied Philosophy, 25* (3), 162–177. See also Chap. 6 for a discussion.

[4]Evidently this is not intended as an exhaustive list.

[5]Unless stated otherwise we have used Irwin's translation of the Nicomachean Ethics. Aristotle (1999). *The Nicomachean Ethics*, (translation and introduction by T. Irwin), Hackett, 2nd edition.

5.1.3 *The Beginnings of Virtue Ethics*[6]

Virtue ethics has its origin in ancient Greece where it was developed by thinkers like Socrates, Plato and Aristotle among others. They were primarily interested in studying and elaborating on the virtue—the driving force—rather than considering the action as such. Very broadly speaking they approached ethics by asking 'what traits of character makes one a good person?', which stands in stark contrast to the core question asked in most modern moral theories, i.e. 'what is the right thing to do?'. Virtue ethics is thus concerned with what kind of persons we should be, what kind of characters we should have, and from that it follows how we should act.

We believe that such an approach has the advantage of being more in line with most people's strong moral intuitions (by this we mean that they are well considered, stable and withstand the test of time). Arguably, a purpose of normative ethics is to help us bring some order to and explain our reflective moral intuitions. Thus it is important to take into consideration that virtue ethics gives a more plausible account of our intuitions. In fact, given the nature of the issues addressed here it is perhaps all the more important to develop an ethical framework, which can help capture the moral concerns of both lay-people and specialists.[7]

Aristotle recognized that ethics is not a science and that it had to be approached differently—the scientific method would not help us in teasing out these moral truths nor capture the essence of ethics. This has led many readers to assume that Aristotle was deeply sceptical about rules, even that he rejected rules all together. Such an interpretation appears unfortunate and it is often used as a basis for claiming that virtue ethics lacks action guiding capacity. It is sometimes said that the theory fails to offer substantive normative advice to vacillating agents and that it is too weak for being a stand-alone normative theory. We disagree with this. In fact it can well be argued that there are a number of rules in virtue ethics, e.g. "always act virtuously" and the virtues themselves. While we do not have the chance to get to the roots of this debate here, we will suggest that (for example) the virtue of *phronesis*, the doctrine of the mean, deliberation and situation sensitivity in virtuous decision-making all contribute to action guidance (see below in Sects. 5.1.4 and 5.1.5). On a more general level, Aristotle recognized that moral decision-making is hard and that it takes a mature moral agent to know what the right thing is when faced with a difficult situation.

---

[6]Parts of section has been previously published in Björkman, B. (2008). Virtue Ethics, Bioethics, and the Ownership of Biological Material.

[7]For some comments on the merits of an inclusive dialogue on these issues see Chap. 6.

5.1.4 What Is the Virtuous Life and What Is so Good About It?[8]

5.1.4.1 Eudaimonia

Aristotle argued that the supreme human good is *eudaimonia*. This is the happy and fulfilled life for any human being and when we lead it.[9] Although he did not say it explicitly it would be fair to assume that Aristotle would have agreed that the desire to lead a fulfilled life is implanted in us by nature. It is far more than just an option among other equally good alternative lives, and in the Nicomachean Ethics he sought to show the reader what kind of a person she needs to be in order to lead this happy life.

On this account *eudaimonia* is the ultimate justification for living in a certain way. It is rational to want *eudaimonia* as Aristotle conceived of it because it is only then we flourish, i.e. realize all our capacities and are fully human. A fulfilled life is not just a set of actions—it is a set of actions performed by someone who does them because she correctly sees the point in doing them.

Moreover, the eudaimon life consists of all intrinsically worthwhile actions and as such it is always the best life available to us. Adding something to such a life will not mean an improvement because it necessarily includes all the activities that are valuable for humans. One way of interpreting this would be the following; it is the theory of happiness that has to be complete and self-sufficient. But even if we accept that *eudaimonia* is the best possible life, that alone does not explain what sort of life it is nor which activities we should engage into fulfil this end. The answer to those questions has to do with our nature and the skills/capacities which are special to humans (man's *ergon*).

Aristotle's whole system is firmly grounded in the study of human nature and human motivation. Very broadly speaking the way to know what to do, according to Aristotle, is to seek the judgment of a good man. Such a good person would know what the right thing to do would be for any agent in a given situation. This is highly relevant as virtue ethics is about being sensitive to situations, to what the circumstances require and then to be motivated to act in the right way. So to know what is good for us we need to know what kind of beings we are. For Aristotle humans, animals and plants all have souls. Not in the sense that they all have a conscious aim (*telos*), but more in the sense that they have an internal organization which explains how they typically behave, that "its organizational purposiveness governs all its activities" (Hughes 2001, p. 34). Even though the types of souls are different (as a result of being organized differently), we all share one thing; the well-being of any organism depends on how well it can exercise its capabilities. It should be noted that *ergon* has to do with activity and expresses itself in action.

---

[8]Parts of Sect. 5.1.4 has been previously published in "Virtue Ethics—a short introduction" in (ed.) Pragati Sahni, *Understanding Ethics, Macmillan India,* 2012.

[9]We use the words 'happy', 'good' and 'fulfilled' as synonyms for the virtuous life.

5.1.4.2 The Function Argument

Aristotle's favoured method for discovering what human fulfilment consists in is called 'The Function Argument'. This is a normative account stipulating that facts about human nature should determine what is good for a human being. In the Nicomachean Ethics the description of the Function Argument is somewhat short but this was not a problem as the reader was assumed to have the necessary background e.g. from having read Plato's Republic. In fact, Aristotle both made a lot of assumptions about his students and had a quite narrowly defined person in mind. Already at the beginning of the teaching the students would subscribe to a certain set of values influencing the way they viewed the world. These values are called the first principles (*archai*) and some examples would be laws of nature and basic intuitions of the kind that *eudaimonia* equals human good. When humans function properly, i.e. when in a state of *eudaimonia*, they are exercising the capacities of the human soul (*ergon*) in a good way. Aristotle writes that if there is more than one such capacity, i.e. that *ergon* is not a singular capacity but a bundle of virtues, then fulfilment is to perform that activity which is the best.[10]

5.1.4.3 The Doctrine of the Mean

Aristotle writes "Virtue, then, is a state that decides, consisting in a mean, the mean relative to us, which is defined by reference to reason, that is to say, to the reason by reference to which the prudent person would define it" (Nicomachean Ethics NE11007a1-a3). Very generally speaking one could describe this as saying that an excellence is an intermediate in an ethical triad framework. Flanked by two vices—one dealing with excessive behaviour and one with deficient—the excellence in the middle issues in actions that neither goes too far nor falls short.

This mean, however, is not an exact point on either scale—the courageous person (for example) hits the mark that is appropriate under the circumstances. Equally, it is important to note that the quantity is not the relevant aspect here. If, for example, one fears ten different things along the lines of great pain, humiliation, poverty and death, then fearing these would be perfectly alright by Aristotle's standards; but if these ten things are different breeds of cats then one is simply wrong. Further to this point it is important to recall that courage (for example) is not a virtue because it is a disposition in the mean nor are cowardice and fearlessness vices because they are deficit and excess. It is the other way around; courage is a virtue because it is the right disposition to fear and therefore it falls in the mean.

So what then is this elusive 'right amount'? Aristotle's idea was that the reason for which a feeling is said to be in the mean (with regards to a number of parameters such as amount, time, intensity, reason and so on) is because that is how a virtuous

[10]Nicomachean Ethics 1098a17–20.

person would feel in that situation.[11] Notably, it is not the case that there is a 'right feeling' floating around out there and can be decided independently of the virtuous agent. That said the right-making feature as such is not simply that it is felt—the underlying reason for why this feeling is correct is of course a story about human nature which is universally true (i.e. the Function Argument).[12]

5.1.4.4 The Virtues

According to Aristotle there are two types of (moral) virtues. First, there are epistemic (or intellectual) virtues—they belong to that part of the soul that has reason.[13] Second, there are virtues of character—they concern the part of the soul that has feelings and desires but which can listen to reason.[14] As the moral virtues have to be acquired through habit and an 'ethical education' it comes down to re-organizing people's desires. It is about setting things straight in the soul as this will ensure that the agent gets pleasure from noble acts and pain from bad.[15] In order to do the right thing in a given situation the agent needs both the motivation (i.e. the desire) to act in that way and the situation sensitivity to see what the right action is in that particular circumstance.

The Greek word *arete* means virtue in the sense of excellence and to be virtuous is to be excellent at doing something. In that sense the virtues are outward and production oriented and they are about actions. Many of the Greeks held that humans should be virtuous for the sake of this life—now is the time when one will be rewarded and as the virtuous action is also the most pleasurable this 'reward' is not simply external. The justification of the moral for the Greek philosophers was 'because it is good for me'; to be moral was to behave in an egoistically rational way.

Aristotle put forward three conditions for an act to count as virtuous. The agent has to (a) have practical knowledge, i.e. know what she is doing, (b) chose the act and chose it for its own sake and (c) the act must flow from a firm character. In other words, a virtue has to be a habitual disposition which gives rise to relatively fixed patterns of behaviour.[16] To assess if an action is virtuous we need to know how the agent saw what she was doing and that it was not down to luck, self-control and so on.

---

[11]Evidently some virtues fit more easily into the doctrine of the mean model than others.

[12]Bostock's suggested solution is that Aristotle should be understood as saying that a virtue is a middling disposition—it is the disposition that the virtue is flowing from that lies in the middle.

[13]For a discussion of the epistemic virtues which plausibly would be conducive to good-decision making see Sect. 5.2.8 below.

[14]For a discussion see /NE1102b30-34/. Sometimes a distinction is made between moral virtues and eudemonic virtues. With regards to Aristotle we believe this to be confusing, for him all moral virtues were eudemonic virtues.

[15]Lear J. (1988). *Aristotle—the desire to understand.* Cambridge University Press, p. 168.

[16]For example brittleness which is a dispositional property of glass and this influences the behaviour of glass when dropped i.e. it shatters.

5.1.5 In Summary

The story told in the Nicomachean Ethics is that if we manage to instill a set of virtues (epistemic and character related) and then act in accordance with them our lives will go infinitely better than would otherwise be the case.[17] The virtues will make us sensitive to the relevant (moral) factors in difficult situations and enable us to deliberate well and reach the type of decisions which are conducive to our long-term well-being (on an individual and collective level). Note, however, that the virtues are not instrumental to the good life—doing the fine and noble is *eudaimonia*.

Broadly speaking we follow Aristotle in his account of the good life and, further, accept the idea that once properly instilled the virtues tend to issue in action. That is, the virtuous agent can in (most) cases be trusted to behave in accordance with virtue.[18] Evidently this is not to say that his account is unproblematic but given its format this volume does not permit further discussion. Instead, we now turn to the question we are interested in investigating in more detail namely: do most people have the cognitive capacities required to instill these virtues and be guided by them in a lasting, generalizable and stable manner? In other words—we accept that this is the good life but is it a life that is available to us?

5.2 Part B: From Meditation to the Good Life

5.2.1 Training for Virtue[19]

On the Aristotelian account (moral) virtue comes in stages, through education and habituation, and it includes both cognitive and emotional dimensions.[20] In his discussions he touches on a question which many modern moral philosophers (and others) struggle with, i.e. can virtue be taught and acquired by practicing or is it

[17]To Aristotle our likes and dislikes indicate whether or not we have acquired the virtues and the virtuous only take pleasure in doing the fine and noble.

[18]As there is no space to go deeper into Aristotle's account of akrasia and moral failure we refer the reader to Book 7 of the Nicomachean Ethics. Aristotle (2002). *The Nicomachean Ethics*, (translation, introduction and commentary by S. Broadie and C. Rowe), Oxford University Press. Burnyeat M. F. (1980). Aristotle on Learning to be Good. In, *Essays On Aristotle's Ethics*, ed. A O. Rorty, 1980, University of California Press, pp. 69–92. Sorabji R. (1980). *Necessity, cause and blame: perspectives on Aristotle's theory*. London.

[19]Some parts of 5.6. have been published in "Virtue Ethics, Bioethics and the Ownership of Biological Material", Björkman B. (2008). Theses in Philosophy from the Royal Institute of Technology 28. viii + 203 pp. Stockholm.

[20]Burnyeat M. F. (1980). Aristotle on Learning to be Good. In, *Essays On Aristotle's Ethics*, ed. A O. Rorty, 1980, University of California Press, p. 71.

perhaps part of human nature?[21] Below follows a very brief account of some key stages in the moral development as Aristotle viewed it.

5.2.1.1 The That

The first step is to develop *the that* (knowledge in a qualified sense) and the second step is to develop *the because* (knowledge in the unqualified sense). When we have both these skills we are able to work out for ourselves how to act in any given situation. We have a deep understanding and have internalized the virtues fully. As a result we desire to do the noble, we take pleasure in doing it and we begin to live the eudaimon life.

In Aristotle's world his students would learn *the that* by listening to his lectures and then, with lots of practice and over time, they would also be able to acquire *the because*. Knowing *the that* means knowing what it is to do the fine and noble, to have a good grasp both of the character virtues and the epistemic virtues (so that one can read the situation right). The student knows these things because she has observed the behaviour of good men and listened to their advice. She has learned by heart like she would any other subject and is slowly beginning to realize how to do this on her own.

Notably this is both hard and time consuming. As Aristotle put it "And those who have just learned something do not yet know it, though they string the words together; for it must grow into them, and this takes time." (Nicomachean Ethics 1147a22-24).

5.2.1.2 The Because

Gradually, as the student becomes familiar with a great number of situations and practice decision-making she begins to internalize the virtues (i.e. become able to know how to act in new situations). This is the beginning of the second step, of knowing *the because*. A mature moral agent knows both *the that* and *the because* and is able to act without effort (i.e. in the sense of self-constraint) and with pleasure. In doing this she meets one of Aristotle's key criteria as a student namely understanding what would be the virtuous choice and taking pleasure in doing just that.[22] In Aristotle's own words "…for actions in accord with the virtues to be done temperately or justly it does not suffice that they themselves have the right qualities. Rather, the agent must also be in the right state when he does them. First, he must know [that he is doing virtuous actions]; second, he must decide on them, and

---

[21]Nicomachean Ethics 1179b20–1180a.

[22]See Dworkin 'Laws & Empire' for a similar analogy. Dworkin R. (1986), *Law's Empire*, Belknap Press.

decide on them for themselves; and, third, he must also do them from a firm and unchanging state." (Nicomachean Ethics 1105a29–1105b1).

In theory then, most of us would be able to (re)train our preferences and develop a taste for the fine and noble. As moral novices, and intermediates, we work towards becoming agents with settled characters who do the virtuous, willingly and for its' own sake.

5.2.1.3 Moral Responsibility

If this is correct, becoming a virtuous person and a good moral decision-maker is mainly about practice, motivation and effort.[23] One consequence of this is that we can be held accountable for our morality or lack thereof, a notion which fits well a common sense idea of responsibility.

In one respect this is of course very appealing—while it might put pressure on the individual it also implies that we are free to shape ourselves in various ways and, consequently, impact how our lives pan out. Clearly such ideas are a good match for the modern (Western) emphasis on the individual (as opposed to the group) and her development.

But what if this is not true—what if we are not quite so free to carve out our future? What if the virtuous life is not a possibility for the many but that the reasons are not primarily socio-economic but a result of biological limitations in the brain? That no matter the amount of education and training a great many will never be able to lead the happy life. If that is the case then it would follow that many people would be excluded from the good life simply because of bad luck in the genetic lottery.[24] Evidently that would have far-reaching implications for moral responsibility and would be hard to square with contemporary ideas about fairness and equality. Indeed it could threaten to undermine virtue ethics as a theory of the best and most complete life.[25]

5.2.2 Creating Moral Experts

A long-standing critique of virtue ethics is that it is too demanding. That for a number of reasons (ranging from socio-economic background to intelligence) the life described by Aristotle as the best is in fact unachievable for the vast majority.

---

[23]Notice that we follow Aristotle's account of moral motivation as broadly understood. E.g. to be morally motivated leads to action, i.e. it is more than simply feeling or thinking that X is right. For more on this see Part A of this chapter.

[24]Notice that we do not mean that socioeconomic and other 'nurture' aspects are not powerful in shaping who we become and what we are able to make of our biological background.

[25]Naturally, one can still accept that it is the eudaimon life but that it might not be a possibility for most people, or indeed, anyone.

While this has been hotly debated amongst philosophers[26] for a long time it seems that recent discoveries in the fields of neuropsychology, moral psychology and even behavioural economics have furnished the critics with new ammunition. As shown in Chaps. 2–4 most people are undermined in their decision-making as a consequence of incomplete cognitive capacities (e.g. various cognitive bias, limited working memory and poor focus/attention). In addition, we have low meta-awareness which results in poor risk assessment, inability to regulate emotions and a reluctance to detach from ourselves (which is needed to access a wider variety of perspectives). In light of such findings it does not seem entirely plausible that the majority of people really have the amount of cognitive skills required to live the good life in a modern society. If that is true then it is hard to see how classical education *alone* would be enough to instil the virtues and turn us into mature moral decision-makers. Such practices might, best case scenario, produce 'moral experts' which is not to be confused with 'true excellence' (which was what Aristotle spoke about). Human excellence is an *unconditional disposition* to act, to feel and generally respond in ways typical of the good person. Expertise, on the other hand, is simply the ability to act and respond (and perhaps in some cases feel) in the ways typical of the sort of expert in question. The main problem with moral expertise, according to Aristotle, is that the person is unreliable. She can have the ability and yet prefer to exercise it badly or not at all. For example she might only use the virtue strategically as a means to an end that wanted to achieve. The virtuous individual, on the other hand, is completely committed and takes pleasure only in the fine and noble.

Still, it might be too soon to abandon the idea that the virtuous life is the happy and good life for humans. Below follows an account of how the mental training and meditation techniques explored in Chaps. 2–4 could, as a result of their positive influence on our cognitive skills (via functional and structural changes), facilitate the habituation process and making virtue ethics an actual and achievable option.

---

[26]For both sides of the discussion see e.g. Harman. 1999. Moral philosophy meets social psychology: virtue ethics and the fundamental attribution error. Proceedings of the Aristotelian Society 99: 315–331. Darley, and Batson. 1973. "From Jerusalem to Jericho": a study of situational and dispositional variables in helping behavior. JPSP 27: 100–108. Doris. 2002. Lack of character. Cambridge: CUP. Haidt, J., Seder, J. P., & Kesebir, S. (2008). Hive psychology, happiness, and public policy. *The Journal of Legal Studies*, *37*(S2), S133–S156. Hursthouse. 1991. Virtue theory and abortion. Philosophy and Public Affairs 20(3): 223–246. Nussbaum. 1986. The fragility of goodness: luck and ethics in greek tragedy and philosophy. UK: CUP. Haidt. 2004. Intuitive ethics; how innately prepared intuitions generate culturally variable virtues. Daedalus 133 (4): 55. Milgram, S. (1963). Behavioral study of obedience. *The Journal of Abnormal and Social Psychology*, *67*(4), 371. Hobbes, T. (1969). *Leviathan, 1651*. Scholar Press.

5.2.3 *Enhancement as an Enabler*

We have sketched an account of how certain types of meditation and mental training techniques can help people to become cognitively enhanced. By improving a set of core cognitive skills we would be better placed to embark on the habituation process and actually have a chance at instilling the virtues. Indeed, a broad uptake of such life-style practices could contribute to enabling more people could overcome (some) of the biological constraints that, no fault of their own, stand between them and good and responsible decision-making.

If correct, this could, help shift the normal (Gaussian) distribution of such cognitive skills to the right. It would benefit those who are on the lower end of the curve by giving them the cognitive tools to manage their lives better and make more well-informed choices. This might, for example help to mitigate and overcome phobia, experiential avoidance and obsessive-compulsive behaviour. Further, it would also benefit the group with capacities which we identify as normal to highly functioning. These would be the great many who have comparatively strong cognitive skills but yet struggle to meet the demands of the modern information society.[27]

This type of training and step-by-step improvements of capacities would plausibly involve introspection and self-reflection. In this process the individual would become more skilled at recognizing flawed thought-patters as well as more motivated (and capable) to change them. This raised awareness links in well with Aristotle's idea of the virtuous agent being capable of explaining her behaviour in hind-sight. He held that one way in which the virtuous person differs from the rest is that she can explain her behaviour if asked. By elaborating on her reasoning she proves that it was more than a lucky guess. But to be able to provide such an explanation would require the agent to be capable of noticing the own behaviour, then detach enough to introspect and analyse when warranted.[28]

5.2.4 *Enhancement as a Way to Increase Commitment*

The cognitive improvements (regarding, for example, cognitive flexibility, meta-awareness, controlled mind-wandering and emotional regulation) could

---

[27]Note that we are primarily interested in a set of core cognitive skills and how they can be conducive to epistemic virtue. In other words we are not looking at a general increase of IQ and what may be the effects of such changes. However, we have identified one study where relational frame training (which is the theoretical basis for Acceptance and commitment therapy) seem to increase the IQ of the participating students: Cassidy, S., Roche, B. & Hayes, S. C. (2011). A relational frame training intervention to raise Intelligence Quotients: A pilot study. *The Psychological Record*, 61, 173–198.

[28]Fleming, S. M., Weil, R. S., Nagy, Z., Dolan, R. J., & Rees, G. (2010). Relating introspective accuracy to individual differences in brain structure. *Science*, *329*(5998), 1541–1543.

contribute to a deeper understanding of the virtuous life as 'the best life' and, thus, a stronger motivation to lead it.

It could work as a self-reinforcement in the sense that the more cognitively skilled the individual, the more motivated and receptive to the idea of keeping up the life-style habits.[29] Further she would be better placed to assess, and more motivated to take up, other life-style changes which might contribute to her epistemic capacities. Such a scenario fits well with the central Aristotelian idea that we ought to commit to this life for its own sake and, further, that we are in it for the long-haul. That we through dedication and effort become a little better and, while we are in the process of change we actually lead the good life.[30]

More speculative but not implausible, this might bring about a positive spin which reinforces the (good) behavior. In other words, the more epistemically virtuous the agent becomes the more motivated will she be to continue the meditative practices, which then will improve her cognitive flexibility further, which is conducive to instilling the virtues (both new ones and deepen existing ones) and so on and so forth.[31] The effects would be beneficial for the agent both as an individual and as a member of collective society. For a good example of how e.g. Character Trait Training potentially could have generalizable effects consider a recent study by Richard Davidsons group,[32] where a mindfulness kindness curriculum was delivered to pre-schoolers, who showed increased cognitive flexibility and delay of gratification and less selfish behaviour, compared to the control group. It appears that their newly found capacity (i) generalized subconsciously and (ii) found a new expression, i.e. went from general kindness to concrete sharing. Similar effects have been shown in adults who have done meditation with elements of compassion training. For example, this resulted in improved perspective taking, self-compassion and decreased levels of perceived stress.[33] Compassionate response to suffering has

[29]Admittedly, research on so called "change processes" indicate that a profound change in a virtuous direction is sometimes the result of a life crisis as opposed to intellectual reasoning. We are also aware of that there is no shortage of examples of individuals groups and organizations which on closer inspection only pay lip service to virtue. However, such empirical observations hardly undermine the foundational assumptions of virtue theory. Plante, Thomas G. and Courtney Daniels. "The Sexual Abuse Crisis in the Roman Catholic Church: What Psychologists and Counselors Should Know". *Pastoral Psychology, Vol. 52, No. 5, May 2004.*

[30]While there is a little means and a lot of ends in all the virtues they are certainly not instrumental to happiness but have substantive intrinsic value.

[31]Notably we are not implying that virtue theory on the traditional account includes moral duty to maximise.

[32]Flook, L., Goldberg, S. B., Pinger, L., & Davidson, R. J. (2015). Promoting prosocial behavior and self-regulatory skills in preschool children through a mindfulness-based kindness curriculum. *Developmental psychology*, *51*(1), 44.

[33]Wallmark, Erik; Safarzadeh, Kousha; Daukantaite, Daiva; Maddux, Rachel E. Promoting Altruism Through Meditation: An 8-Week Randomized Controlled Pilot Study. Mindfulness, September 2013, Vol. 4(3), pp. 223–234.

also been shown after meditation training, such as offering ones chair for a disabled person (i.e. overcoming the bystander effect).[34]

5.2.5 *Helpful but not Enough*

Not implausibly then, improved core capacities like cognitive flexibility could be highly conducive to the *instilling* of epistemic virtues such as introspection, intellectual courage, intellectual honesty and impartiality. In addition it would also make it easier for the agent to *remain committed* to the virtuous life as she would be better placed to assess it from an all things considered perspective. As a result, agents might choose to act in ways which are (more) pro-social as their virtues would issue in a set of broad life-skills that would make us, for example, less selfish, biased and short-sighted.[35]

While this is positive it does not solve the problem. Firstly, the documented effects on cognition and behaviour are not powerful enough on their own to bring about the type of improvement that we seem to need (see Chaps. 2 and 3). Secondly, improved cognitive skills do not guarantee good moral behaviour. Recall, for example, the previous discussion on the moral expert, i.e. the person who uses the virtues as instruments to secure strategic advantages. Consequently we need a moral framework within which the cognitive capacities will be put to good use.

In way of summary: we have assumed then that the techniques for cognitive enhancement explored here could enable more people to both to understand the point of, and commit to, the virtuous life. Hence, making the mental training techniques described here an integral part of the habituation process could increase the likelihood of actually instilling the virtues. In addition, as we become more epistemically virtuous we will be able to capitalize further on the structural and

[34]Condon, P., Desbordes, G., Miller, W. B., & DeSteno, D. (2013). Meditation increases compassionate responses to suffering. *Psychological science*, *24*(10), 2125–2127.

[35]As regards the truth tracking capacities of the epistemic virtues we subscribe to a Reliabilist account as broadly conceived of i.e. that the epistemic virtues include motivating and reliability components which means that they can help us attain the truth or, at least, help us attain more true beliefs than false ones. Under such a broad umbrella assumption the epistemic virtues can be defined as "dispositions to attain the truth and avoid error in a certain field of propositions F, in certain conditions C." (Sosa E., *Knowledge in Perspective*, 138, 141). Some scholars have added various constraints many of which are internalistic. As summarised by Heather Battaly "…Greco has maintained that the intellectual virtues must be both reliable and grounded in the subject's conformance to the epistemic norms that she countenances. Plantinga's properly functioning faculties are reliable, but they also operate in accordance with a design plan in an environment sufficiently similar to that for which they were designed. And, Zagzebski maintains that the virtues have a motivational component in addition to a reliability, or success, component. She thinks the virtues are enduring, acquired excellences of a person that involve the motivation for truth and reliable success in attaining that end of that motivation." Battaly's paper '*What is Virtue Epistemology?*' https://www.bu.edu/wcp/Papers/Valu/ValuBatt.htm#top.

functional changes achieved through the (continuous) mental training and meditation. Such practices could help many to level up to a point where they have a realistic chance to embark on the habituation process. But while this may both prime and groom us it is a facilitator for—not a replacement of—epistemic virtue.

5.2.5.1 Why We Need the (Epistemic) Virtues[36]

While cognitive enhancement could make important contributions it would be a mistake to grant such techniques the status of an alternative strategy for the good life. Two arguments support this claim. The first argument focuses on the epistemic and moral capacities that the agent develops, while the second argument is concerned with the intrinsic value of the process itself.[37]

Firstly, it is likely that the capacities the agent ends up with through the type of cognitive enhancements that are available today, are not on par with the overall situation sensitivity and capacity for skilful deliberation which is achieved by instilling the virtues. Being virtuous means being sensitive to contexts and situations in a very fine-tuned way. It involves a substantial element of sound judgment as well as properly directed sentiments, and it allows the agent to be highly discriminating when she exercises the virtues and responds to situations. The virtues transform the agent into a stable and reliable decision-maker who knows what to do in the 'all things considered' sense. This, in turn, enables her to be flexible and thus well equipped to face the changing reality and ever-increasing flow of information in society. Capacities like sound judgement and situation sensitivity are also likely to continuously re-enforce the agent's own commitment to virtue, thus minimising the development of enhanced but immoral agents. Moreover, such abilities will be helpful when evaluating the potential goodness of emerging enhancement technologies.

By choosing the virtuous life, the agent would not only be better off from an all-things-considered aspect, but also with regards to the development of the individual virtues. In other words, she would be good both at getting the information right and at making the decisions.

Indeed, in being virtuous the agent will be better both at concrete decision making and knowing when, what kind and to whom epistemic deference would be

---

[36]Section 5.2.5.1 has been previously published in B. Froding, Cognitive Enhancement, Virtue Ethics and the Good Life, *Neuroethics*, (2011) 4:223–234.

[37]We acknowledge that forms of very advanced conative (i.e. behavior directed toward action) enhancement potentially might achieve both the same results and mimic the experience of habituation. This text however, deals with the type of medical and technological possibilities that we have access to today or are likely to have in the near future. For space reasons, this book cannot deal with conative enhancement as a separate issue but for an interesting argument, see e.g. Douglas (2008) 'Moral enhancement', *Journal of Applied Philosophy, Vol. 25, No. 3*: Persson and Savulescu (2008) 'The Perils of Cognitive Enhancement and the Urgent Imperative to Enhance the Moral Character of Humanity', *Journal of Applied Philosophy, Vol. 25, No. 3.*

appropriate.[38] In light of the medical and technological knowledge we have today, it appears improbable that cognitive enhancements would be able to rival both this general sense of equity and the individual virtues. Another aspect (see Chap. 6) is that agents who habitually act virtuously could (as a beneficial side-effect) be likely to bring about a society where the institutions subscribe to a set of institutional virtues, for example transparency.[39]

Secondly, cognitive enhancements are unlikely to mimic all the worthwhile aspects of the virtuous life. Consider, for example, the intrinsically valuable process of habituation that agents are expected to undertake. What matters are not only the capacities the agent hopefully manages to develop at the end of the process, but also the experience of acquiring and exercising the virtues. One of the most central features of virtue ethics is that doing the virtuous thing is leading the good life. In other words, the actual process is valuable in itself; it is a key part of *eudaimonia* also as the agent goes through it.[40]

Arguably then, those interested in becoming well-informed, stable and reasonable decision-makers we have good reason to seek to develop and strengthen a wide range of epistemic virtues. Below follows some examples of such epistemic virtues and how they in a more concrete way would be conducive to good decision-making.

5.2.6 The Role of Epistemic Virtues

5.2.6.1 To Think Well

According to Aristotle a hallmark of good decisions is that they are guided by reason as opposed to desire. This is not to say that rational decision-making is void of emotion—but rather that we are not slaves to unreflected desire and that we have learned to attach pleasure to the right type of actions.

As moral decision-making is very hard we need to have skills (or virtues) which make us flexible and functional (as opposed to rigid and over controlling or reliant on a set of rules which cannot grasp the complexity but will lead us to

---

[38]Here we follow the Aristotelian account in assuming some version of cognitivism i.e. that the virtuous person is the one who knows what is right and wrong.

[39]For some interesting ideas on social moral epistemology, see e.g. Buchanan. 2007. Institutions, beliefs and ethics: Eugenics as a case study. Journal of Political Philosophy 15/1: 22–45. Buchanan. 2009. Philosophy and public policy: a role for social moral epistemology. Journal of Applied Philosophy 26(3): 276–290.

[40]For an account of the temporal aspects of different virtues and personal goods, i.e. the idea that certain virtues are good for us at certain points in our lives (for example, that innocence and trustingness is good for children but less so for adults), see Slote. 1983. Goods and virtues. New York: Clarendon. Note, however, that Slote does not claim that all virtues are 'relative' in this sense.

over-simplifications). In order to become such a reliable and mature moral decision maker we need to know ourselves and identify our weaknesses. We need to learn how to assess situations correctly; this involves being sensitive to the relevant salient features of the situation, including myself and my role. Further, we must have the will to act.

As mentioned above the virtues can be split into two groups—the epistemic (or intellectual) virtues and the virtues of character. The epistemic virtues make us excellent at thinking about how we should act whereas the character virtues concern excellence in doing or acting. Together these virtues enable the agent to e.g. take pleasure in the right things, be sensitive to the (relevant) salient features in a situation, deliberate well, choose the right option and then act on it. All the virtues contribute to good conduct in their own way and when they are combined they help us assess worthwhile goals and the means to reach them.[41]

5.2.6.2 The Intellectual Virtues According to Aristotle

In the Nicomachean Ethics Aristotle presents five separate intellectual virtues. He explains that while each has its own task they are not equally important and thus ought to be considered in a special order.[42] They are: scientific knowledge (*episteme*), craft knowledge (*techne*), practical wisdom (*phronesis*), intellect (*nous*) and last but certainly not least there is wisdom (*sophia*).

Although not ranked the highest of all the intellectual virtues *phronesis*, or practical wisdom, merits a few extra comments given our context. This capacity enables us to direct and harmonize the different virtues in order for them to form a coherent whole. Consequently, *phronesis* is much more than 'the right principle'. It is in fact a dynamic virtue which involves both good deliberation and cleverness or intuition.

To have *phronesis* means to be skilled at thinking about how one should act in order to live a worthwhile life. Such an individual is good at thinking morally, i.e. she both knows the moral principles and how to apply them in practical situations. There are some important similarities between *phronesis* and cognitive flexibility. For example, to be cognitively flexible means that the individual has an increased awareness and understanding of how she should live. In practice this can mean that she has the capacity to handle both internal and external (sometimes moral) challenges which might hinder this preferred life style.

---

[41] The relationship between the intellectual and the emotional in moral decision-making is among the most disputed issues in the Nicomachean Ethics. For examples of contradictory account see e.g. Book 3.2–3 and compare it with Book 6.12–13 in the Nicomachean Ethics.

[42] Pakaluk M. (2005). *Aristotle's Nicomachean Ethics*. CUP. p. 208.

5.2.7 *Some Modern or New Epistemic Virtues that Might Be Useful*

While we need all the virtues (to what extent is disputed) to lead the good life and flourish, it is the epistemic virtues which enable us to identify the alternatives and then to ponder them well.

So while we are likely to benefit from having the 'old' virtues to a higher degree, they might not suffice in the face of a changing society. In order to be the type of good and responsible decision-makers envisioned here we might require an extended set of virtues. These new virtues would enable people to assimilate information, form beliefs and, possibly, even act in a more epistemically conscientious and impartial way.

Seeking to understand and improve the ways in which humans form beliefs is far from a novel philosophical problem. A fairly recent strategy, however, is the one presented here, i.e. to combine ideas from the areas of virtue ethics and Virtue Epistemology with findings from the natural sciences. Plausibly, this approach can generate insights as to how the ways in which individuals and societies make important decisions can be improved. At this stage a short introduction to Virtue Epistemology might be in order. The field of virtue epistemology is young but interest is growing rapidly both from a theoretical and a more applied perspective.

In the last two decades a lot of work is being done trying to identify some cognitive virtues which might prove essential for a happy life in a modern society. Very broadly speaking there are two camps; the virtue Reliabilists[43] and the virtue Responsibilists.[44] Both groups research the type of capacities that enables us to respond well to information and increases our epistemic ability broadly conceived of. As for *exactly* which virtues one might require in order to respond well to information well there is, however, widespread disagreement. The Reliabilists focus on natural, as well as, acquired reliable virtues (or competencies) such as sensory capacities, induction, deduction and memory. The Responsibilists, on the other hand, focus on acquired stable character traits such as open-mindedness and intellectual courage and do not view natural faculties like vision and memory as epistemic virtues proper.[45]

---

[43]See, for example, Axtell (ed.) (2000), *Knowledge, Belief and Character*, (Lanham, MD: Rowman & Littlefield); Sosa (1991) *Knowledge in Perspective*, (New York: CUP); Greco (2000), *Putting sceptics in their place* (New York: CUP).

[44]For example, Zagzebski (1996), *Virtues of the Mind: an inquiry into the nature of virtue and the ethical foundations of knowledge* (New York: CUP); Montmarquet (1993), *Epistemic Virtue and Doxastic Responsibility* (Lanham, MD, Rowman & Littlefield).

[45]This paragraphs has been previously published in Chap. 5 of Fröding, B. (2013). *Virtue ethics and human enhancement*. Springer.

Leaving the details of that particular debate aside, we suggest that the following skills would be good examples of new, or non-traditional, virtues: intellectual honesty, intellectual courage, open-mindedness, tolerance, impartiality, commitment to fairness, capacity for introspection and detachment and improved memory (both with regards to process memory and long-term memory).[46]

Presumably, such capacities are highly conducive to responsible decision-making, improved risk-assessment and could form a counter-weight to the bias and motivated reasoning we tend to suffer.[47] For some concrete examples of how epistemic virtues like intellectual courage, open-mindedness and impartiality could be reflected in improved behavior, consider how such and other, epistemic virtues could reduce or compensate for the various bias we suffer and which many of us are unaware of. While it is not clear that we can impact the actual neurophysiological process in the brain as a way to reduce bias it is likely that stronger epistemic virtues could enable us to balance the negative effects. Ideally, we would be able to nip it in the bud so to speak but barring that modifying the fallout of e.g. rigid normative thinking. Plausibly, epistemic virtues can also improve capacity for accurate risk assessment as we would suffer less miss-placed fears, both in the sense that we fear the right things and to the right extent. Further, all of these examples could make an important contribution to the developing and reinforcement of a sense of equity.[48] This is the capacity we need to make the right choice i.e. 'an all things considered' judgment of how to act in a situation.

Recent studies have shown that cognitive flexibility increases during childhood, and actually is higher in adolescents than in adults.[49] If it is correct that our cognitive flexibility decreases with age we would have even stronger reasons to engage in training techniques which might go some way to maintain, best-case scenario increase, this and other core capacities. In addition, we might also have extra incentive to cultivate the epistemic virtues as they might compensate (with regards to e.g. decision-making) for some of the aging related deterioration.

[46]Evidently this is not intended as a complete list.

[47]By motivated reasoning we mean decision-making based on unreflected gut reactions NOT people's strong moral intuitions which would be well considered, stable and able to withstand the test of time. We define a bias here as a form of heuristic or shortcut that the human brain is prone to when engaging in e.g. decision-making, general assessment of events, ranking how important events/facts are and what to pay attention to in a situation. Unfortunately the agent tends to be unaware of the nature and magnitude of such bias.

[48]For more on equity see Book 5.9 of the Nicomachean Ethics.

[49]Hauser TU, Iannaccone R, Walitza S, Brandeis D, Brem S. Cognitive flexibility in adolescence: neural and behavioral mechanisms of reward prediction error processing in adaptive decision making during development. Neuroimage. 2015 Jan 1;104:347–54.

5.2.8 *From Core Capacities to Decision-Making*

This section provides a tentative account of how the core cognitive capacities previously identified can be connected to (more) responsible decision-making.[50]

5.2.8.1 From Physiological Changes in the Brain to Daily Matters—Functional Aspects

In Chaps. 2–4 we described methods for improving a set of core cognitive and emotional regulation skills e.g. cognitive flexibility, increased focused attention (meta-awareness), 'controlled' mind-wandering and improved capacity for emotional regulation.

For concreteness we discussed which practical skills such improvements could be cashed out as. We proposed that cognitive flexibility could issue in; less misplaced fear, fewer bias (better capacity to compensate and/or control bias) and heightened creativity (without getting into detail we imagine that this would involve a strong capacity for both divergent and convergent thinking). Increased meta-awareness in combination with emotional regulation could issue in e.g. a capacity for compassion (both with others and oneself),[51] emotional stability, stress-resilience, equanimity, self-detachment and introspection.[52]

Arguably, these skills would constitute functional improvements in the daily lives of the individual. For another example, consider how a capacity to step-back and detach from oneself might provide another perspective. This could enable the agent to view both herself, others and general problems (e.g. global warming, political and religious conflict) in a different, perhaps less entrenched and more solution orientated, light. Similarly, a reduction of (misplaced) fear would presumably boosts open-mindedness and might balance some of the negative effects of in-group bias.

As pointed out previously it is important to be able to harbor conflicting emotions and thoughts at the same time without spiraling into a 'systemic collapse'. To have that skill, i.e. to be able to stand the cognitive dissonance[53] and resist the

---

[50]Although this text is inspired by a number of Aristotelian ideas, it features a mixed set of virtues. On a general note, however, we believe that the central claim defended here is Aristotelian in spirit.

[51]For a discussion on the self-regarding and other-regarding aspects of virtue ethics see Fröding, B. (2010). On the importance of treating oneself well. *Polish Journal of Philosophy*, *4*(1), 7–21. For a much broader discussion on compassion, empathy and over-coming the self-other duality see Coplan, A., & Goldie, P. (Eds.). (2011). *Empathy: Philosophical and psychological perspectives*. Oxford University Press.

[52]I.e. in the epistemically virtuous manner, as opposed to a cold, calculating strategy or as a self-harming strategy to avoid discomfort.

[53]Cognitive dissonance is described as the mental stress or discomfort experienced due to two or more contradictory beliefs, ideas, or values at the same time, or a confrontation by new

temptation to jump to a conclusion[54] just to reduce stress in the brain has a clear bearing on how well our lives go and to what extent we can pursue and achieve the goals that are valuable to us in life.

It seems quite plausible that (everything else being equal) an individual who can focus on the relevant parameters in a situation and regulate her own emotions to a higher degree than her peers would be better at deliberating.[55] If she also is able to notice the level of understanding and emotional balance in the others, i.e. their "mindset", and take those factors into account while deliberating, that would on the one hand increase the complexity of the deliberation process and on the other also hopefully lead to an all things considered better outcome. For such multifactorial deliberation a good working memory and meta-awareness would be key capacities. Similarly, the capacity to zoom in (with sustained focused attention) and out (with open awareness), and to know when to do what, is another key competence that is increasingly needed. To use moments of mind-wandering in a controlled and "primed" way, when pondering moral decisions would plausibly be conducive to improved decision-making.

The examples given above remind us of the sense of equity and the ability to make the 'all things considered judgements' which Aristotle talked about. A sense of equity is in many ways a hallmark of the virtuous agent and on the Aristotelian account it is connected to a sense of justice. Someone who has equity knows how to interpret the spirit of the law and avoids both making mistakes and becoming overly rigid by following it by the letter. She can adapt to the circumstances without losing sight of what is truly just in a situation. This deep understanding for what it really means to be just coupled with her situation sensitivity enables her to read even very complicated situations right. In a modern context we might call her a truly decent person or a highly functional individual.

These are the skills of the mature moral decision-maker. The person who has instilled the virtues and has a robust yet flexible ability to pick up on the relevant salient features in a situation, to deliberate them and to identify the action(s) which would be virtuous. Arguably then, this is a good example of how the improvements of core cognitive capacities could promote the development of epistemic virtues which then issue in concrete choices and behaviour.

(Footnote 53 continued)

information that conflicts with existing beliefs, ideas, or values, in one individual (it could also be manifested in groups or at a societal level). Festinger, L. (1957). *A Theory of Cognitive Dissonance*. California: Stanford University Press. Festinger, L. (1962). "Cognitive dissonance". *Scientific American* 207 (4): 93–107.

[54]See also Sect. 1.5.2 we tend to perceive meaningful patterns in random data (*apophenia*) and to perceive vague and random images and sound as significant (*pareidolia*), not least human faces. Voss, J. L., Federmeier, K. D., & Paller, K. A. (2011). The potato chip really does look like Elvis! Neural hallmarks of conceptual processing associated with finding novel shapes subjectively meaningful. *Cerebral Cortex*, bhr315.

[55]For a discussion on the role of choice and the concept of parity in virtue theory see Fröding, B., & Peterson, M. (2012). Virtuous Choice and Parity. *Ethical theory and moral practice*, *15*(1), 71–82.

5.2.9 Commonalities and Compatibility

At this stage it might not come as a complete surprise that classical meditation and virtue ethics overlap in some important assumptions and training techniques. None the less we would like to highlight some such cases as it could be taken as further indication of their compatibility and the idea that they might fortify the effects of the other.

Commonality 1—Priming and grooming the (sub)conscious Meditation is frequently described as the journey inwards, a deep reflection which enables us to observe ourselves, our actions and thoughts.[56] On such traditional accounts the focus is on getting to know thyself and to experience an awakening or enlightenment. The vast majority of such reports are from a 1st person perspective i.e. accounts of subjective experiences which are hard to verify.

All the more interesting then, some recent studies have provided scientific evidence (i.e. objective verification) of physiological changes in the brains and bodies of individuals who meditate regularly. Examples range from lower blood-pressure to changes in the brain which is reflected in improved cognitive capacities and reduced levels of stress and anxiety. Such studies might indicate that this vague 'spiritual awakening' which has been reported from a first person perspective for millennia, is in fact a very concrete wake up call for the brain e.g. with regards to pathways, synapses, structures and functions.[57]

A lot of our decisions are made sub-consciously but during those processes we are none the less capable of some kind of reasoning. In order for that reasoning to lead to beneficial decisions however, it needs to be properly trained and 'groomed'.[58] Such training can come in the form of past experiences, including, for example, meditation and mental training. This is especially useful when time is scares as it enables the brain to reach a beneficial decision much quicker. One can also work with the outcome of the decisions, and focus on behavioral changes one aspires to. While we still cannot fully control the decision-making processes we can improve the chances of the outcome being one which is in line with our valued direction. We can prime/groom these 'subconscious' decision-making processes into different "directions" as broadly conceived and a concrete way of doing just that is to practice virtuous deliberation in similar scenarios.

Similar ideas can be found in virtue ethics. Consider, for example, how the education process comes down to a lengthy priming or grooming of the mind

---

[56]Citation: The longest journey is the journey inward. Hammarskjöld D (2006) Markings, transk W.H. Auden, Leif Sjoberg (NY Vintage Books).

[57]E.g. Britton, W. B., Lindahl, J. R., Cahn, B. R., Davis, J. H., & Goldman, R. E. (2014). Awakening is not a metaphor: the effects of Buddhist meditation practices on basic wakefulness. *Annals of the New York Academy of Sciences*, *1307*(1), 64–81. For more details on the neurophysiological background please see Chap. 2 in this book.

[58]See pervious discussion of the work of Kahneman in Chap. 3. See also Damasio, A. (2012). *Self comes to mind: Constructing the conscious brain*. Vintage.

(i.e. more than just adjusting one's general preferences). As explained in 5.6 the student learns 'the that' and then 'the because'—through great reflection (and practical training) becomes a person of equity. Someone able to think, respond and feel in ways typical of the virtuous agent. Once habituated the agent is firmly committed to this way of life.

Commonality 2—Training both inward and outward focus A contributing reason for focusing on the mental training techniques and meditation is that (some) such techniques purposefully train both outward focus and inward focus.

It is well documented that outward focus can be trained through mediation practices.[59] Indeed, a state of open monitoring has also been described from a neurophysiological perspective.[60] Similarly, inward focus can be cultivated by meditation and mindfulness techniques such as body-scan, which increases introceptive awareness.[61]

This can then be contrasted with a discussion in virtue ethics regarding the self-regarding and other-regarding aspects of the virtues. Traditionally, it has been common for virtue ethicists to label the character virtues as either other-regarding or self-regarding.[62] Based on this division, the former category is said to be primarily concerned with the benefit to others, whereas the main concern of the latter would be the benefit to the agent herself.[63]

An alternative interpretation would be that virtues like friendship, justice, even temper and generosity, all seen as other-regarding on the above account, in fact contain central self-regarding aspects.[64] And, further, that these aspects are in fact so important that an agent cannot be said to possess the particular virtue to the full extent unless she displays them in action. She must act virtuously both with regards to others and to herself, and in order to qualify as fully virtuous, an agent would

[59]Outward focus can also be trained efficiently via specialized computer games and we do not wish to imply that these, and other, life-style practices could not be combined for better effect. For comments see Chap. 6).

[60]Manna A, Raffone A, Perrucci MG, Nardo D, Ferretti A, Tartaro A, Londei A, Del Gratta C, Belardinelli MO, Romani GL. Neural correlates of focused attention and cognitive monitoring in meditation. Brain Res Bull. 2010 Apr 29;82(1–2):46–56.

[61]Farb NA, Segal ZV, Anderson AK. Mindfulness meditation training alters cortical representations of interoceptive attention. Soc Cogn Affect Neurosci. 2013 Jan; 8(1):15–26.

[62]For an interesting account see e.g. Slote M. (1992), From Morality to Virtue, OUP and Taylor G. & Wolfram S. (1968). The Self-regarding and Other-regarding Virtues, *The Philosophical Quarterly*, 18(72): 238–248. For a traditional account see e.g. Mill J. S (1997) 'On Liberty', Ryan A. (ed.), New York, Norton or e.g. book 3, Chap. 9.3 in Sidgwick H. (1981), *The methods of ethics, 7th edition*. Indianapolis: Hackett Publishing Company, and von Wright G. H (1963). *The varieties of goodness*, London: Routledge & Kegan Paul, e.g. p. 153.

[63]This is not addressing the issue of who actually benefits from the virtuous action. For an interesting discussion on this see e.g. Foot P., Moral Beliefs, in Proceedings of the Aristotelian Society (1958–59), 59:83–104, Foot P. (1978). Virtues and Vices and Other Essays in Moral Philosophy. Berkeley: University of California Press; Oxford: Blackwell.

[64]See Fröding, B. (2010). On the importance of treating oneself well. *Polish Journal of Philosophy*, *4*(1), 7–21.

have to satisfy both the other-regarding and the self-regarding aspects of these virtues.[65] Slightly more formally put, the idea is that the exercise of character virtue X with respect to oneself (in the correct circumstances) is a necessary, but not sufficient, condition for the full possession of said character virtue. This dual capacity including outward focus and accurate assessment as well as introspection brings to mind a current discussion regarding the delicate balance between compassion towards others and self-compassion in the field of e.g. social psychology and neuro-economics.[66] Plausibly, having the capacity both for reflecting on the self and one's context, as well as a more general propensity for reflective thinking is central to developing the virtues. The fact that such abilities can be trained and improved through the mental training and meditation techniques presented here is of interest.

Commonality 3—Boosting life-skills Somewhat sweepingly, it could be said that both virtue ethics and meditation practices are about becoming 'good at living'. It is about gaining a deeper understanding for what to attach importance to and to engage in a special way of thinking.

When meditators practice they often describe that they lose track of time. One theory of what actually happens from a neurophysiological perspective when they have this experience is this;

The human brain always strives for solution and coherence. When we find a solution the internal reward systems are activated and we feel rather good. Unfortunately this is not to be enjoyed for long. Almost immediately the brain starts looking for the next solution and the next and the next. In meditation the brain is 'tricked' into thinking that it has found a solution, but since there is no content (i.e. no solution to ponder and then move on from) the brain can engage in a state of flow. Perhaps one can say that the brain gets stuck in the reward phase.

Admittedly, this can be achieved by a range of activities e.g. running, playing music, listening to music and is not unique to meditation. Indeed, it can be said that meditation is just another tool—a vehicle to achieve harmony and balance in the brain. None the less it appears that Attention Focus Meditation is a comparatively efficient way to become better at resisting mental conflict which usually occurs as soon as we are exposed to competing stimuli. It seems to help us both (i) get better at ignoring irrelevant stimuli and (ii) get better at regaining attention and focus on the relevant features of the situation.

Flow described as 'lack of mental conflict' is quite similar to the state of harmony and pleasure and well-being experienced by the virtuous agent. She knows that this is the best life overall and does not suffer inner conflict and various cognitive dissonance. Meditation could, not implausibly, facilitate this state by enabling the agent to develop the equity and *phroneis* which is required for mature and reliable moral decision-making.

---

[65]This argument presupposes that one regards the virtues as threshold concepts, a view which we take to be relatively uncontroversial and so will not defend here.

[66]Neff, K. (2011). *Self compassion*. Hachette UK.

Related to this, consider the similarities between Character Trait Training and the sensitivity of salient features which Aristotle mentions. Character Trait Training helps to cultivate certain capacities, for example noticing suffering in one self and others, and act in order to decrease this suffering and enhance wellbeing.[67] Improving and stabilizing (getting the balance right between 'other' and 'self' for one thing) such traits has much in common with the development of equity and that fine-tuned sensitivity which is required for mature and reliable moral decision-making.

5.2.10 Conclusion

This chapter has provided an account of (i) how the epistemic virtues are connected to virtue ethics and (ii) how and why some such traits are good for us from both a rational and moral point of view, i.e. that they are desirable because they make our lives go better. Examples of useful epistemic virtues include; intellectual honesty, intellectual courage, tolerance, impartiality, open-mindedness, commitment to fairness, capacity for introspection and detachment and improved memory (both with regards to process memory and long-term memory). We have argued that such cognitive skills are highly conducive to responsible decision-making.

While recognizing that the virtuous life was an acquired taste and that it might seem very demanding to many people, Aristotle maintained that it was the best and most complete life. In the Nicomachean Ethics he explained that the way to instil the virtues is a combination of education, habituation and a lifelong commitment. Just like other forms of knowledge the virtues will come in stages and the process is likely to be long and require considerable effort and dedication. The goal is to acquire the unconditional disposition to act, feel and respond in the way that the virtuous person would in the situation.[68]

A number of findings in modern neuroscience paint a bleaker picture and calls into question if the virtuous life is achievable at all. It has been shown that most people are undermined in their decision-making as a consequence of low cognitive capacities (e.g. various cognitive bias, limited working memory and poor focus/attention). As if that was not enough, humans tend to have low meta-awareness which results in poor risk-assessment, inability to regulate emotions and an unwillingness to detach from ourselves. If the neuroscientists are right

[67]For a discussion see Chap. 2 of this book. For a review on compassion, see Goetz, J. L., Keltner, D., & Simon-Thomas, E. (2010). Compassion: an evolutionary analysis and empirical review. *Psychological bulletin*, *136*(3), 351.

[68]For a good discussion on 'dispositions in ethics' see Part IV Ethics and Epistemology in Groff, R., & Greco, J. (Eds.). (2013). *Powers and capacities in philosophy: the new Aristotelianism*. Routledge.

and the best explanation for our low cognitive capacities and lack of virtue is found in biology then attempting to instil stable values and character traits in the young (and old) by way of education and training seems naïve at best and utterly futile at worse. For one thing, such limited cognitive capacities can hardly be conducive to the type of stable character traits and 'all things considered moral judgements' that Aristotle called for.

We have attempted to sketch a way in which life style choice and classical training can be combined. The mental training and meditation techniques contribute by enhancing the agent to a cognitive functioning where she both encompasses the point of, and has a chance to embark on the habituation as described by Aristotle. Plausibly, improved cognitive flexibility perhaps especially in combination with emotional regulation, gives a better understanding of the various problems at hand and this, in turn, would issue in the individual developing a set of epistemic virtues which would guide her decision-making. Increased capacity to develop and maintain certain character traits (intellectual and moral virtues) would then, in part, be a consequence of improved brain capacity. As previously explained we recognize that the question of motivation is very important when talking about life style choice and *change* but, for space reasons, we cannot go into that debate here.

In addition, such a combination might increase the appeal of virtue ethics by making the good life more achievable and less dependent on luck, which, in turn, fits very well with the central virtue ethics idea that agents are responsible for their morality, or lack thereof.

The next chapter makes some concluding remarks on the role of society and the importance of combining the forms of training described here with other methods (not in the least traditional education) in order to increase the chances of actual results.

References

Aristotle. (1999). *The Nicomachean ethics*, (trans. and introduction by T. Irwin) (2nd Edn.). Indianapolis: Hackett.

Hughes, G. (2001). *Routledge philosophy guidebook to Aristotle on ethics*. New York: Routledge.

Chapter 6
Conclusions

Abstract On providing a brief summary of the arguments presented in the book Chap. 6 turns to discuss the potential for combining a wider range of life-style practices for greater effects. Examples include classical education, physical as well as mental training and playing specialized computer games. A brief resume of Aristotle's account of *eudaimonia* including the idea that agents who are successful in instilling virtues plausibly can be taken to lead happier lives than those who fail to do so is presented. This Chapter also comments on the potential of embedding structures and how such measures might incentivize more pro-social behavior.

Keywords Mental training · Life-style · Virtue · Decision-making · Pro-social · Neuroenhancement

6.1 Summary

Over the past five chapters we have discussed how a number of measurable structural and functional changes in the brain can be achieved through various life-style practices and training methods. It appears that most adult humans are able to create new neurophysiological connections, as well as to some extent also new functions—and consequently new behaviors—through training activities they engage in with some regularity.

Very broadly speaking we have argued that meditation training enables us to influence some aspects of our biological make-up and could, for example, boost our cognitive flexibility as well as our ability (and propensity) to act compassionately. More in particular, we have looked at the beneficial effects of Attention Training Meditation, Open Monitoring (which is included in e.g. Mindfulness Stress

B. Fröding and W. Osika, *Neuroenhancement: How Mental Training and Meditation Can Promote Epistemic Virtue*, SpringerBriefs in Ethics,
DOI 10.1007/978-3-319-23517-2_6

Reduction techniques) and Compassion Training.[1] While these are not the only life-style practices with proven effects on the brain we find them especially interesting as they can improve a number of core cognitive capacities which, in turn, play a central role for people's quality of life and general well-being. Our examples include; cognitive flexibility, focus/attention and meta-awareness, controlled mind-wandering and emotional regulation.

We then proceeded to sketch an account of how a number of such cognitive changes might improve the capacity for instilling and maintaining a range of character traits (primarily epistemic virtues) as identified by Aristotle and some modern virtue epistemologists. Our examples of new, or non-traditional, virtues include; intellectual honesty, intellectual courage, open-mindedness, tolerance, impartiality, commitment to fairness, capacity for introspection and detachment and improved memory (both with regards to process memory and long-term memory). For an example of how such virtues can work together consider that in order to be 'good at remembering' the individual also has to be *willing* to remember even when it might be painful or embarrassing. Another example of capacities 'collaborating' are situations where one has to be cognitively able to zoom in on, but also willing to listen to, signals of stress (which often are displaced) and to reappraise what is going on. Such 'willingness' would be promoted by e.g. intellectual honesty, courage, introspection and self-detachment.

Having such epistemic virtues to a higher degree than what is currently the case for most people, could issue in improved deliberation and moral decision-making. It might, for example, form a counter-weight to the bias and motivated reasoning we tend to suffer. Further, it could make an important contribution to the ability both to develop and maintain a sense of equity. This is the capacity we need to make the right choice i.e. 'an all things considered' judgment of how to act in a situation.[2]

Although not explicitly argued here, we follow Aristotle's account of *eudaimonia* including the idea that agents who are successful in instilling such and other virtues plausibly can be taken to lead happier lives (from an all things considered perspective) than those who fail to do so. But even accepting that and aspiring to such a life, it is quite clear that moral decision-making is often very hard. Further, while the effects of the training exemplified here are both generalizable and stable, they cannot be said to achieve **radical** cognitive improvements.

With this in mind, it might be prudent to opt for a combination of (non-rivalrous) training techniques.[3] We will now turn to take a brief look at some possible combinations of practices including e.g. classical education, physical as well as

---

[1]Brefzcynski-Lewis JA, Lutz A, Schaefer HS, et al. (2007). Neural correlates of attentional expertise in long-term meditation practitioners. *Proc Natl Acad Sci.* 104(27):11483–8. Kerr CE, Josyula K, Littenberg R. (2011). Developing an observing attitude: an analysis of meditation diaries in an MBSR clinical trial. *Clin Psychol Psychother.* 18(1):80–93. Gilbert, P. (2010). Compassion focused therapy: Distinctive features. New York, NY: Routledge.

[2]For more on equity see Book 5.9 of the Nicomachean Ethics.

[3]Note that while training might be 'the virtuous thing to do' we do not wish to advance the claim that it would be morally compulsory.

mental training and playing specialized computer games.[4] In addition, we will discuss the potential for embedding structures which might incentivize more pro-social behavior.

6.2 Combinations

6.2.1 *Meeting the Challenge*

At the start of this book we made the case that humanity as a collective need to start making better decisions.[5] We briefly described what we believe to be a rather pressing situation; a combination of a changing society and research confirming that our cognitive capacities are more limited, and our bias more substantive, than what was previously known. Taken together, these factors conspire to undermine the quality of our decision-making which, in turn, negatively impacts our quality of life.[6]

As previously explained (Chaps. 2 and 4) we can prime and groom the (mental) processes which influence behavior and decision-making by engaging in Attention Training Meditation, Open Monitoring and Compassion Training.[7] However, while the cognitive improvements are stable and (in many cases) generalizable outside the training scenario, there is no denying that this is a slow process.

In light of this, it might be wise to engage several methods for cognitive improvements on a regular basis. By this we primarily mean different life-style choices (as opposed to pharmacological cognitive enhancers[8] and technology). As explained in Chap. 3 our position is that given the current scientific knowledge (i.e. a limited understanding of the brain, the cognitive processes and the long-term effects of biomedical enhancers), our best bet for lasting cognitive enhancement will

[4]See Chap. 3 for the full discussion.

[5]Broadly understood as; responsible, reflected and rational decision-making.

[6]See Chap. 1.

[7]Britton, W. B., Lindahl, J. R., Cahn, B. R., Davis, J. H., & Goldman, R. E. (2014). Awakening is not a metaphor: the effects of Buddhist meditation practices on basic wakefulness. *Annals of the New York Academy of Sciences*, *1307*(1), 64–81. Ricard, M., Lutz, A., & Davidson, R. J. (2014). Mind of the Meditator. *Scientific American*, *311*(5), 38–45. Hasenkamp, W., Wilson-Mendenhall, C. D., Duncan, E., & Barsalou, L. W. (2012). Mind wandering and attention during focused meditation: a fine-grained temporal analysis of fluctuating cognitive states. *Neuroimage*, *59*(1), 750–760. Gilbert, P. (2010). Compassion focused therapy: Distinctive features. New York, NY: Routledge.

Klimecki, O. M., Leiberg, S., Ricard, M., & Singer, T. (2013). Differential pattern of functional brain plasticity after compassion and empathy training. *Social cognitive and affective neuroscience*, nst060.

[8]Pharmacological cognitive enhancers, or PCEs, are used to treat neurodegenerative (e.g. Alzheimer's and dementia) and neuropsychiatric disorders (e.g. schizophrenia and ADHD).

come through 'life-style' broadly conceived of.[9] In this we include, for example, meditation, computer games, other relevant online activities, traditional education, physical exercise, diet and supplements.[10] These, and other methods, can often be combined to greater effect. For example, specialized computer games could work well as a complementary tool for improving cognitive skills.[11] Further, we do not wish to sideline the role of traditional education including, for example, specially-designed training packages for both children and adults.[12] Indeed, intentional education even in its classical form is a very powerful and dynamic tool which should not be underestimated in these discussions.

In a recent report issued by Stanford University a group of researchers concluded that "People can cultivate their cognitive powers by leading physically active, intellectually challenging and socially engaged lives."[13] While we agree with this, we certainly do not rule out the future potential of alternative methods (such as pharmaceutical drugs, hormones, neurotransmitters, wearable technology and brain machine interface technology). As mentioned in Chap. 3 the combination of imaging, neurophysiological monitoring and/or treatment modalities such as neuroenhancers, meditation practices and neurofeedback—i.e. multimodal neuroenhancement—could be more powerful than the use of any single method. Admittedly, employing multiple technologies together may add to the expense, time and complexity of studies and treatment protocols. However, if the results of cost-benefit analyses show that the added benefits outweigh the costs, then multimodal neuroenhancement may still be preferred.[14] An especially interesting combination that has been studied is physical exercise with concomitant mental training.[15]

---

[9]As should be clear by now we have nothing against biomedical enhancement in *principal*. Rather, our position is that it (at least on its own) is a lesser option today, and in the foreseeable future.

[10]Evidently much more empirical research as to the effects, their generalizability and how long they last, is required.

[11]See e.g. Clark, V. P., & Parasuraman, R. (2014). Neuroenhancement: enhancing brain and mind in health and in disease. *Neuroimage*, *85*, 889–894. Granic, I., Lobel, A., & Engels, R. C. (2014). The benefits of playing video games. *American Psychologist*, *69*(1), 66.

[12]For interesting examples of how reading fictional literature in an organized way has improved social capacity, empathy and other life -skills see e.g. Kidd, D. C., & Castano, E. (2013). Reading literary fiction improves theory of mind. *Science*, *342*(6156), 377–380. McLellan, M. F., & Jones, A. H. (1996). Why literature and medicine?. *The Lancet*, *348*(9020), 109–111. Billington, J. (2011). 'Reading for Life': Prison Reading Groups in Practice and Theory. *Critical Survey*, *23* (3), 67–85.

[13]Statement issued on Oct 14, 2014 by The Stanford Center for Longevity and the Max Planck Institute for Human Development.

[14]Clark, V. P., & Parasuraman, R. (2014). Neuroenhancement: enhancing brain and mind in health and in disease. *Neuroimage*, *85*, 889–894. Evidently such cost-benefit models must take into consideration the ethical aspects that attach.

[15]Curlik, D. M., & Shors, T. J. (2013). Training your brain: Do mental and physical (MAP) training enhance cognition through the process of neurogenesis in the hippocampus?. *Neuropharmacology*, *64*, 506–514.

While much more research is required, it seems highly likely that we in the not too distant future will see important breakthroughs in the use of pharmacological cognitive enhancers,[16] hormones, neurotransmitters and nootropics, as well as technology based enhancers.[17]

6.3 Moral Enhancement—Different Takes

6.3.1 Why Do We Need the Virtues?

In this book we have explored cognitive enhancement, through lifestyle changes, as a means to moral improvement.[18] While the goal is improve one's decision-making (to become more able to make responsible, reflected and rational decisions and then act on them) and instill a set of epistemic virtues, our model is quite different from what is usually meant by moral enhancement. Consequently a clarification might be in order. When talking about moral enhancement many philosophers refer to biomedical enhancement in order to make us more humane and increase our pro-social dispositions (e.g. more altruistic and empathic and less aggressive and xenophobic).[19]

We, on the other hand, have insisted that if the aim is to improve moral decision-making, the type of cognitive improvement that might be brought about courtesy of biomedical enhancers (of the type we have access to today and in the foreseeable future) is a lesser option than our suggested combination of life-style and virtue ethics. We have advanced three main arguments in favor of this position. Firstly, we also need better cognitive skills to compute all the information, get better at risk assessment, reduce bias etc. hence the moral improvement would not be

[16]Pharmacological cognitive enhancers, or PCEs, are used to treat neurodegenerative (e.g. Alzheimer's and dementia) and neuropsychiatric disorders (e.g. schizophrenia and ADHD).

[17]As for the potential for further acceleration of technological development consider Moore's Law. Plausibly such scenarios would offer both new/increased challenges as well as solutions.

[18]Broadly speaking enhancement can be split into three different types based on the primary area of improvement; physical, cognitive and moral.

[19]For an introduction to the debate see e.g. Persson, I., & Savulescu, J. (2014). Reply to commentators on Unfit for the Future. *Journal of medical ethics*, medethics-2013. Persson, I., & Savulescu, J. (2014). Should moral bioenhancement be compulsory? Reply to Vojin Rakic. *Journal of medical ethics*, *40*(4), 251–252. Carter, J. A., & Gordon, E. C. (2014). On cognitive and moral enhancement: A reply to Savulescu and Persson. *Bioethics. Persson I. & Savulescu J. Getting Moral Enhancement Right. Bioethics 2013;* ***27***(3): 124–131. Persson, I., & Savulescu, J. (2010). Moral transhumanism. *Journal of Medicine and Philosophy*, *35*(6), 656–669. Douglas, T. (2008). Moral enhancement. *Journal of applied philosophy*, *25*(3), 228–245. Shook, J. R. (2012). Neuroethics and the possible types of moral enhancement. *AJoB Neuroscience*, *3*(4), 3–14. Sylvia, T., Guy, K., Sarah, M., Julian, S., Neil, L., Miles, H., & Cowen, P. J. (2013). Beta adrenergic blockade reduces utilitarian judgement. *Biological psychology*, *92*(2), 323–328.

enough to do the whole job (see Chap. 3). Secondly, virtue ethics in combination with life-style changes of the kind we recommend here is a (more) reliable and safe way to achieve moral improvement. Further, even if the biomedical option would be safe the habituation process (see Chap. 5) is in itself valuable as a part of the good life.[20] Thirdly, in addition to being potentially harmful, many biomedical treatments are likely to be expensive and not readily available for the general population. The mental training and meditation techniques described here on the other hand, are relatively cheap (also with regards to opportunity cost as most people can learn the basic methods relatively fast[21]) and could be defended on an equality and fairness basis—not only socio-economic but (presumably) also with regards to intelligence and talents and skills. On that last note, we now turn to look at some ways in which we can promote a certain type of behaviour through social measures. Could, for example, embedding structures facilitate the habituation process and promote pro-social behaviour?[22]

6.4 The Role of Embedding Structures

6.4.1 Virtue Ethics and Social Organisation

Virtue ethics is primarily occupied with what kind of person one ought to be and does not (at least not in its modern form) easily lend itself to clear-cut theories for social organisation. The belief that humans ought to lead virtuous lives does not automatically imply that the surrounding state should seek to bring about such behaviour.[23] Then again, it is hard to deny that there is at least some connection between 'how easy it is to lead the virtuous life' and how the society around one is organized. Evidently society can make it 'easier' or 'harder' for the citizens to act in

[20]For the full argument see Fröding, B. E. E. (2011). Cognitive enhancement, virtue ethics and the good life. *Neuroethics*, *4*(3), 223–234.

[21]Zeidan, F., Johnson, S. K., Diamond, B. J., David, Z., & Goolkasian, P. (2010). Mindfulness meditation improves cognition: evidence of brief mental training. *Consciousness and cognition*, *19* (2), 597–605. Boettcher, J., Åström, V., Påhlsson, D., Schenström, O., Andersson, G., & Carlbring, P. (2014). Internet-based mindfulness treatment for anxiety disorders: A randomized controlled trial. *Behavior therapy*, *45*(2), 241–253.

[22]Notably this is not an attempt to argue that society as a whole ought to be organized in such a way that it promotes, or encourages, virtuous behavior. We do, however, believe that there are sensible ways to construct a *eudaemonia*-based view of social organization.

[23]According to Aristotle, however, there is most certainly such a connection. "It is evident that the best politeia is that arrangement according to which anyone whatsoever might do best and live a flourishing life." Aristotle (1997). *The Politics*, (Politics: books VII and VIII/translated with a commentary by Richard Kraut), Oxford: Clarendon press. 1323a 14–19.

certain ways.[24] Below follows a few comments on the role of embedding structures and their role as possible promoters of virtue (beyond the individual to the surrounding society).

6.4.2 *Cultivating Life-Skills*

We have proposed that most people through a combination of cognitive improvements, better emotional regulation skills and commitment to the virtuous life could acquire a set of 'life-skills' which all things considered are likely to both enable them (and potentially those around them) to fare better in life.

Plausibly, individuals with access to a broad(er) skill-set would be better placed to recognize, and more willing to seek, solutions to some of the pressing global challenges we are facing.

It appears that being more cognitively flexible could bring about a host of such 'generally improved life-skills' e.g. being more adaptable, better at risk assessment, better at emotional regulation and more impartial. Further, it could improve our capacity for epistemic modesty and epistemic deference, i.e. identifying who to trust and defer judgment too on issue that we ourselves have no or little expertise in.[25] Such and other skills could be manifested in our behavior and the choices we make and in fact strengthen our chances to lead autonomous lives.[26] Given how little we know of the future it appears a good strategy to seek to acquire a broad, flexible skill-set as to maximize versatility. Society is growing increasingly complex and this has implications both for what it means to have a good life and to be a good citizen. Indeed, seeking to develop such skills would promote the autonomy and extend the freedom of the individual. Reconnecting to the discussion on moral enhancement above, it deserves pointing out that the step from using biomedical enhancers to strengthen dispositions to those dispositions actually issuing in behaviour, is not always clear. Now, contrast this with the virtue ethics idea of developing a wide range of epistemic and moral virtues which will be reflected in our deliberation and in our actions. While it might of course be questioned how stable those traits actually are it would at least seem a prudent and, further, more versatile choice if the goal is to improve actual moral decision-making.

---

[24]This has been discussed in e.g. different areas of social psychology, including the famous experiments by Zimbardo and Milgram. Zimbardo, P.G. (2007). *The Lucifer Effect: Understanding How Good People Turn Evil*. New York: Random House. Milgram, Stanley (1963). "Behavioral Study of Obedience". *Journal of Abnormal and Social Psychology*, *67*(4): 371–378.

[25]For a good discussion on how to trust wisely see e.g. Levy, N. (2006). Open-mindedness and the duty to gather evidence. *Public Affairs Quarterly*, 55–66.

[26]See Fröding B. and Juth N., *Cognitive enhancement and the principle of need*. Currently under review.

Plausibly, being willing and able to engage in pro-social behaviour and responsible decision-making[27] would be central to what can broadly be considered 'good citizenship'. Some examples of pro-social behavior could include a raised sense of responsibility for matters that fall in the collective domain,[28] a willingness to adopt a sustainable life-style, and other 'living together' aspects as broadly conceived of.

The notion of a wide set of life-skills also links in well with the idea that there are many good lives to be had, i.e. the sort of pluralistic take on worthwhile pursuits which is an integral part of the mature democracy.[29] Connecting this thought to virtue ethics we would like to point to the general sense in which it can be said that there are several ways of life (*bioi*). In the Nicomachean Ethics Aristotle discusses what he considered the strongest candidates for the good life. Unfortunately, the ambiguity in the text has caused a massive, and occasionally heated, debate.

From Book 1 and more or less all the way to Book 10 of the Nicomachean Ethics Aristotle appears to promote an inclusive doctrine of *eudaimonia*. But then towards the end of Book 10.7 he somewhat unexpectedly concludes that rather than being both a practical and a theoretical activity *eudaimonia* is only *Theoria*, i.e. theoretical or contemplative thought. Broadly speaking two very different answers to the question 'what does human fulfilment consist in' or 'why do we do the things we do' are championed; The inclusivist (also called the comprehensive) account and the exclusivist account. The two views yield very different visions of who the truly virtuous being is. Is it someone who both lives a fulfilled life and contributes to society (helps to run things) in a way that perhaps even enables others to lead morally admirable lives? Or is it a person in an ivory tower who sits in isolation and contemplates *Theoria*?

A reasonable conclusion is that, for most people, the good life is a mix between the philosophical and the political.[30] In practice the good life is more than theoretical contemplation, it is also about acting in accordance with the other virtues and participating in society. Consequently, the happy and pleasurable life is likely to involve a mix of politics (i.e. practice) and philosophy (i.e. theory) and it is about getting the balance right.[31] So in actuality there can be many different versions of

[27]As previously noted 'good decision-making' it could, at least in the context of this book, be understood as 'responsible, reflected and rational decision-making'.

[28]Avoiding the tragedy of the commons scenarios as described by e.g. Hardin, G. (1968). The tragedy of the commons. *Science*, *162*(3859), 1243–1248.

[29]For a discussion on reasonable pluralism see Rawls, J. (2005). *Political liberalism*. Columbia University Press, USA. Reflective equilibriums and the role of intuitions see e.g. Brun, G. (2014). Reflective Equilibrium Without Intuitions?. *Ethical Theory and Moral Practice*, *17*(2), 237–252.

[30]For a discussion on what Aristotle considered different kinds of good lives, The Politics, Book 4.1.

[31]To Aristotle our likes and dislikes indicate whether or not we have acquired the virtues to the full extent and the virtuous only take pleasure in doing the fine and noble. This might sound strange as the good life can involve pain and death and loss but a reasonable reading might be that although virtues such as courage might bring great physical pain, doing the right thing means achieving one's goals and is thus still pleasurable.

the happy life and *eudamonia* consists in a package of worthwhile things and activities. One both needs to conduct one's life well and interact in society and be able to stand by and reflect on oneself, life and the world and see how it fits in with the bigger picture. The integration of some of the training techniques suggested in this book in everyday life, could be conducive to more people being able to achieve such mixes.

In addition to individual commitment and practice, the development of life-skills might be facilitated by the creation of 'embedding structures'. Some might argue that it would not only be helpful, but actually necessary, for such frameworks to be in place. Examples include; a social and political system that rewards certain behaviour (e.g. generosity, self-lessness), structural (e.g. reflected in city planning and architecture) and specialized education packages on secular moral development. Perhaps schools should be more active in offering an environment where children more actively could practice and improve their moral decision-making skills. We imagine then that the programs would seek to train both the instrumental ability (i.e. cognitive and motor skills) and provide content (i.e. a set of moral values). One, to us, promising way of beginning to address many such issues would be through the framework of virtue ethics.

Without engaging in the wider discussion on the legitimacy of various types of embedding structures and nudging policies in a given society,[32] it appears that such measures can impact people's behavior both in the shorter and longer run. But efficiency and compliance are, of course, but two parameters. A host of ethical aspects and potential value conflicts immediately present themselves, e.g. autonomy, transparency, freedom of choice and human rights. It goes without saying that they must be carefully considered and balanced. It deserves pointing out, however, that on the virtue ethics account the virtuous life is the best life for the individual who is experiencing it. Consequently, to create an environment which involves both intrinsic and extrinsic motivations could be defended on *other grounds* than paternalism.

6.4.3 Creating an Inclusive Dialogue

Over the past chapters we have tried to show that virtue ethics might facilitate a deeper insight to some of the more pressing ethical challenges attaching to

[32]There are some potential similarities in enhancing the instilling of the virtues we discuss here, and e.g. the discussion about and implementation of physical activity, which has received an increasing interest in society, both from the medical community as well as from the educational system.

Vuori, I. M., Lavie, C. J., & Blair, S. N. (2013, December). Physical activity promotion in the health care system. In *Mayo Clinic Proceedings* (Vol. 88, No. 12, pp. 1446–1461). Elsevier. Babey, S. H., Wu, S., & Cohen, D. (2014). How can schools help youth increase physical activity? An economic analysis comparing school-based programs. *Preventive medicine*, *69*, S55–S60.

enhancement in general and cognitive enhancement in particular. An advantage when compared with alternative theories is that virtue ethics provide a richer and more complex understanding of the nature of the fulfilled life for human.

In addition, as it tends to be (more) in line with most people's moral intuitions (of the strong stable kind) it can assist in the creation of an inclusive, open and informed dialogue involving laypeople and scientists and politicians.[33] This dialogue ought to be cross-cultural and inter-religious. Evidently the purpose is not simply to validate and justify what people happen to think. Our point is that we need to create an inclusive dialogue where citizens are invited to actively discuss and reflect together. This would take, on the one hand, an active and engaged citizen who partakes in society and, on the other hand, a transparent society that promotes open and inclusive dialogue with, and among, the citizens.[34]

Arguably, such open dialogue could be contributing to the type of coordinated governance necessary to meet the many environmental and developmental challenges which, on a global scale, are threatening to greatly undermine our levels of well-being. For some examples of such threats to our future prosperity and security consider the annual Global Risk Report issued by the World Economic Forum. Major risks identified in recent reports include; income disparity, extreme weather events, fiscal crisis and technology risks like cyberattacks and critical information infrastructure breakdown.[35] These are scenarios considered capable of causing systemic shock on a global scale (for more on such challenges see Chap. 1). For a practical example from the enhancement debate, consider the developments in fields such as technology and medicine. As individuals and as a collective we need to ponder both *what type* of cognitive enhancements we should allow as well as *which techniques* we should use. The new discoveries hold great promise but as science leaps forward the ethical aspects that attach also grow ever more challenging.[36] To flourish in such a society we need not only improved cognitive capacities but also a robust moral fabric which is action guiding and conducive to deep analysis as well as inclusive dialogue.[37]

In a recent study involving some 4000 participants Fitz et al. explored public attitudes towards cognitive enhancement. One conclusion was that "The data collected suggests that the public is sensitive enough to and capable of understanding the four cardinal concerns identified by neuroethicists, and tend to cautiously accept

---

[33]Note that the type of intuitions that we have in mind here is not the fast, instinctive, unreflected, gut-driven and automatic thinking often described as System 1 thinking in the literature. For more on this see Chap. 1 and the discussion on Kahneman and Twersky.

[34]See e.g. Global Risks 2014 Report from the Davos meeting http://www.weforum.org/reports/global-risks-2014-report.

[35]http://www.weforum.org/reports/global-risks-2014-report.

[36]For some rather dystopic reflections on such challenges see e.g. Byung-Chul Han, (2013) *Im Schwarm*. Matthes & Seitz Verlag, Berlin.

[37]Evidently we are not implying that there is no such discussion today but, rather, that it can become better informed and more productive and solution orientated.

cognitive enhancement even as they recognize its potential perils."[38] This seems to speak in favour of the type of inclusive, society wide dialogue on neuroethics in general and cognitive enhancement in particular, that we have defended in this book. In addition to public understanding being "sufficiently sophisticated to merit inclusion"[39] one might argue that (at least in a welfare based healthcare system) the average tax payer ought to have a say and, further, that an open dialogue is an efficient way to expose the type of urban myths, misinformation, conspiracy theories and outright lies, but also inflated optimism, that tend to attach to the development of new technologies and medicine. Regardless of ones position on the actual use of cognitive enhancements minimizing misplaced fears appears to be desirable.

Further, this call for an on-going dialogue sits well with the virtue ethics idea that the moral and intellectual virtues need to be practiced in order to be exercised well. In addition—given the rate a change in society we would need to continue to develop the virtues in order to maintain that detailed sensitivity which is key to all things considered judgements.

6.5 Concluding Remarks

6.5.1 Beyond Rules

Aristotle believed that a complete view of how one should behave cannot be successfully codified. As ethics is not a science to treat it as such would be to miss the whole philosophical point. The question 'how should we live?' cannot be given a codifiable answer beyond the recommendation to acquire the virtues and only take pleasure in doing the fine and noble. To him that was the wrong approach—as rules could not possibly capture all potential scenarios they will end up becoming a hindrance and an over-simplification rather than help. Moral decision-making is instead about learning how to become sensitive to circumstance, to see what is right to do in a situation as this is the only reliable skill we can hope for, the only one versatile enough. The mature moral decision-maker achieves this sensitivity through a combination of practical wisdom, the moral and intellectual virtues and a sense of equity.

Evidently, developing this level of situation sensitivity is no mean feat and, as previously explained, virtue theory is often criticised for painting a (near) impossible picture of the good life. Indeed, many virtue ethicists would recognise that it is a great challenge, although a worthwhile one.

---

[38]Fitz, N. S., Nadler, R., Manogaran, P., Chong, E. W., & Reiner, P. B. (2013). Public attitudes toward cognitive enhancement. *Neuroethics*, 1–16.

[39]Fitz, N. S., Nadler, R., Manogaran, P., Chong, E. W., & Reiner, P. B. (2013). Public attitudes toward cognitive enhancement. *Neuroethics*, 1–16.

Hence it is not implausible that even those who find Aristotle's account of the good life and habituation process convincing, could be open to some cognitive enhancement methods. In light of (a) the pressing situation and (b) the growing body of solid studies from the natural science indicating that we are less cognitive skilled than previously thought—even bona fide virtue ethicists might conclude that providing people with tools for better emotional regulation and lifting them up to a cognitive level where they have a chance to begin to embark on the instilling of the virtues might be a good thing.

As for the choice of method it appears that given the relatively small cognitive improvements that can be achieved through the methods which meet the criteria of being safe and available today, it would be prudent to opt for a combination. We imagine that such packages, in addition to the mental training and meditation techniques described here, would include traditional education and a potentially wide range of life-style habits (including diet and cardio vascular exercises). We have also suggested that it might be of interest to further explore the role of social embedding strategies in order to encourage certain types of behavior.

It is worth pointing out that virtue theory contains strong internal embedding elements. The more (epistemically) virtuous one becomes the deeper the understanding of why the good life (as described by Aristotle) is, all things considered, a better option than the alternatives. This re-enforces the commitment and might bring about a virtuous loop or circle. It could become a positive spin which leads to a heightened motivation. Plausibly, this process is facilitated by the type of generalizable capacity[40] to make 'better all things considered choices' which can be strengthened through meditation. This capacity would, once the epistemic virtues are instilled, extend to various domains in life and, in turn, provide a strong incentive to keep up the training. Last, but certainly not least, anyone seeking to become a (more) conscientious epistemic agent also ought to cultivate a sense of epistemic modesty.

[40]Note that the capacity we are primarily interested in is cognitive flexibility and how that can be conducive to epistemic virtue. In other words we are not looking at a general increase of IQ and what may be the effects of such changes. However, we have identified one study where relational frame training (which is the theoretical basis for Acceptance and commitment therapy) seem to increase the IQ of the participating students: Cassidy, S., Roche, B. & Hayes, S. C. (2011). A relational frame training intervention to raise Intelligence Quotients: A pilot study. *The Psychological Record*, 61, 173–198.

25278639R10071

Made in the USA
San Bernardino, CA
23 October 2015